www.EffortlessMath.com

... So Much More Online!

✓ FREE Math lessons

✓ More Math learning books!

✓ Mathematics Worksheets

✓ Online Math Tutors

Need a PDF version of this book?

Please visit www.EffortlessMath.com

ISEE Middle Level Mathematics Prep 2019

A Comprehensive Review and Ultimate Guide to the ISEE Middle Level Math Test

By

Reza Nazari

& Ava Ross

Copyright © 2018

Reza Nazari & Ava Ross

All inquiries should be addressed to:

info@effortlessMath.com

www.EffortlessMath.com

ISBN-13: 978-1-970036-05-3

ISBN-10: 1-970036-05-2

Published by: Effortless Math Education

www.EffortlessMath.com

Description

ISEE Middle Level Mathematics Prep 2019 provides students with the confidence and math skills they need to succeed on the ISEE Middle Level Math, building a solid foundation of basic Math topics with abundant exercises for each topic. It is designed to address the needs of ISEE Middle Level test takers who must have a working knowledge of basic Math.

This comprehensive book with over 2,500 sample questions and 2 complete ISEE Middle Level tests is all you need to fully prepare for the ISEE Middle Level Math. It will help you learn everything you need to ace the math section of the ISEE Middle Level.

Effortless Math unique study program provides you with an in-depth focus on the math portion of the exam, helping you master the math skills that students find the most troublesome.

This book contains most common sample questions that are most likely to appear in the mathematics section of the ISEE Middle Level.

Inside the pages of this comprehensive ISEE Middle Level Math book, students can learn basic math operations in a structured manner with a complete study program to help them understand essential math skills. It also has many exciting features, including:

- Dynamic design and easy-to-follow activities
- A fun, interactive and concrete learning process
- Targeted, skill-building practices
- Fun exercises that build confidence
- Math topics are grouped by category, so you can focus on the topics you struggle on
- All solutions for the exercises are included, so you will always find the answers
- 2 Complete ISEE Middle Level Math Practice Tests that reflect the format and question types on ISEE Middle Level

ISEE Middle Level Mathematics 2019 is an incredibly useful tool for those who want to review all topics being covered on the ISEE Middle Level test. It efficiently and effectively reinforces learning outcomes through engaging questions and repeated practice, helping you to quickly master basic Math skills.

About the Author

Reza Nazari is the author of more than 100 Math learning books including:
– **Math and Critical Thinking Challenges:** For the Middle and High School Student
– **GED Math in 30 Days**
– **ASVAB Math Workbook 2018 - 2019**
– **Effortless Math Education Workbooks**
– **and many more Mathematics books …**

Reza is also an experienced Math instructor and a test–prep expert who has been tutoring students since 2008. Reza is the founder of Effortless Math Education, a tutoring company that has helped many students raise their standardized test scores—and attend the colleges of their dreams. Reza provides an individualized custom learning plan and the personalized attention that makes a difference in how students view math.

You can contact Reza via email at:
reza@EffortlessMath.com

Find Reza's professional profile at:
goo.gl/zoC9rJ

Contents

Chapter 1: Arithmetic

Topics that you'll learn in this chapter:

- ✓ Simplifying Fractions
- ✓ Adding and Subtracting Fractions
- ✓ Multiplying and Dividing Fractions
- ✓ Adding Mixed Numbers
- ✓ Subtract Mixed Numbers
- ✓ Multiplying Mixed Numbers
- ✓ Dividing Mixed Numbers
- ✓ Comparing Decimals
- ✓ Rounding Decimals
- ✓ Adding and Subtracting Decimals
- ✓ Multiplying and Dividing Decimals
- ✓ Converting Between Fractions, Decimals and Mixed Numbers
- ✓ Factoring Numbers
- ✓ Greatest Common Factor
- ✓ Least Common Multiple
- ✓ Divisibility Rules

Simplifying Fractions

| *Helpful* *Hints* | – Evenly divide both the top and bottom of the fraction by 2, 3, 5, 7, ... etc.
– Continue until you can't go any further. | **Example:**

$\frac{4}{12} = \frac{2}{6} = \frac{1}{3}$ |

✎ *Simplify the fractions.*

1) $\frac{22}{36}$

2) $\frac{8}{10}$

3) $\frac{12}{18}$

4) $\frac{6}{8}$

5) $\frac{13}{39}$

6) $\frac{5}{20}$

7) $\frac{16}{36}$

8) $\frac{18}{36}$

9) $\frac{20}{50}$

10) $\frac{6}{54}$

11) $\frac{45}{81}$

12) $\frac{21}{28}$

13) $\frac{35}{56}$

14) $\frac{52}{64}$

15) $\frac{13}{65}$

16) $\frac{44}{77}$

17) $\frac{21}{42}$

18) $\frac{15}{36}$

19) $\frac{9}{24}$

20) $\frac{20}{80}$

21) $\frac{25}{45}$

Adding and Subtracting Fractions

Helpful	– For "like" fractions (fractions with the same denominator), add or subtract the numerators and write the answer over the common denominator.
Hints	– Find equivalent fractions with the same denominator before you can add or subtract fractions with different denominators.

– Adding and Subtracting with the same denominator:

$$\frac{a}{b} + \frac{c}{b} = \frac{a+c}{b}$$
$$\frac{a}{b} - \frac{c}{b} = \frac{a-c}{b}$$

– Adding and Subtracting fractions with different denominators:

$$\frac{a}{b} + \frac{c}{d} = \frac{ad+cb}{bd}$$
$$\frac{a}{b} - \frac{c}{d} = \frac{ad-cb}{bd}$$

✏️ Add fractions.

1) $\frac{2}{3} + \frac{1}{2}$

2) $\frac{3}{5} + \frac{1}{3}$

3) $\frac{5}{6} + \frac{1}{2}$

4) $\frac{7}{4} + \frac{5}{9}$

5) $\frac{2}{5} + \frac{1}{5}$

6) $\frac{3}{7} + \frac{1}{2}$

7) $\frac{3}{4} + \frac{2}{5}$

8) $\frac{2}{3} + \frac{1}{5}$

9) $\frac{16}{25} + \frac{3}{5}$

✏️ Subtract fractions.

10) $\frac{4}{5} - \frac{2}{5}$

11) $\frac{3}{5} - \frac{2}{7}$

12) $\frac{1}{2} - \frac{1}{3}$

13) $\frac{8}{9} - \frac{3}{5}$

14) $\frac{3}{7} - \frac{3}{14}$

15) $\frac{4}{15} - \frac{1}{10}$

16) $\frac{3}{4} - \frac{13}{18}$

17) $\frac{5}{8} - \frac{2}{5}$

18) $\frac{1}{2} - \frac{1}{9}$

Multiplying and Dividing Fractions

Helpful *Hints*	– **Multiplying fractions:** multiply the top numbers and multiply the bottom numbers. – **Dividing fractions:** Keep, Change, Flip Keep first fraction, change division sign to multiplication, and flip the numerator and denominator of the second fraction. Then, solve!	**Example:** $\dfrac{a}{b} \times \dfrac{c}{d} = \dfrac{a \times c}{b \times d}$ $\dfrac{a}{b} \div \dfrac{c}{d} = \dfrac{a}{b} \times \dfrac{d}{c} = \dfrac{ad}{bc}$

✎ Multiplying fractions. Then simplify.

1) $\dfrac{1}{5} \times \dfrac{2}{3}$

4) $\dfrac{3}{8} \times \dfrac{1}{3}$

7) $\dfrac{2}{3} \times \dfrac{3}{8}$

2) $\dfrac{3}{4} \times \dfrac{2}{3}$

5) $\dfrac{3}{5} \times \dfrac{2}{5}$

8) $\dfrac{1}{4} \times \dfrac{1}{3}$

3) $\dfrac{2}{5} \times \dfrac{3}{7}$

6) $\dfrac{7}{9} \times \dfrac{1}{3}$

9) $\dfrac{5}{7} \times \dfrac{7}{12}$

✎ Dividing fractions.

10) $\dfrac{2}{9} \div \dfrac{1}{4}$

13) $\dfrac{11}{14} \div \dfrac{1}{10}$

16) $\dfrac{3}{5} \div \dfrac{1}{5}$

11) $\dfrac{1}{2} \div \dfrac{1}{3}$

14) $\dfrac{3}{5} \div \dfrac{5}{9}$

17) $\dfrac{12}{21} \div \dfrac{3}{7}$

12) $\dfrac{6}{11} \div \dfrac{3}{4}$

15) $\dfrac{1}{2} \div \dfrac{1}{2}$

18) $\dfrac{5}{14} \div \dfrac{9}{10}$

Adding Mixed Numbers

Helpful *Hints*	Use the following steps for both adding and subtracting mixed numbers. **Example:** − Find the Least Common Denominator (LCD) − Find the equivalent fractions for each mixed number. − Add fractions after finding common denominator. − Write your answer in lowest terms.

Centered example:

$$1\frac{3}{4} + 2\frac{3}{8} = 4\frac{1}{8}$$

✏️ *Add.*

1) $4\frac{1}{2} + 5\frac{1}{2}$

2) $2\frac{3}{8} + 3\frac{1}{8}$

3) $6\frac{1}{5} + 3\frac{2}{5}$

4) $1\frac{1}{3} + 2\frac{2}{3}$

5) $5\frac{1}{6} + 5\frac{1}{2}$

6) $3\frac{1}{3} + 1\frac{1}{3}$

7) $1\frac{10}{11} + 1\frac{1}{3}$

8) $2\frac{3}{6} + 1\frac{1}{2}$

9) $5\frac{3}{5} + 5\frac{1}{5}$

10) $7 + \frac{1}{5}$

11) $1\frac{5}{7} + \frac{1}{3}$

12) $2\frac{1}{4} + 1\frac{1}{2}$

Subtract Mixed Numbers

Helpful	Use the following steps for both adding and subtracting mixed numbers.	**Example:**
Hints	Find the Least Common Denominator (LCD) – Find the equivalent fractions for each mixed number. – Add or subtract fractions after finding common denominator. – Write your answer in lowest terms.	$5\frac{2}{3} - 3\frac{2}{7} = 2\frac{8}{21}$

✎ **Subtract.**

1) $4\frac{1}{2} - 3\frac{1}{2}$

2) $3\frac{3}{8} - 3\frac{1}{8}$

3) $6\frac{3}{5} - 5\frac{1}{5}$

4) $2\frac{1}{3} - 1\frac{2}{3}$

5) $6\frac{1}{6} - 5\frac{1}{2}$

6) $3\frac{1}{3} - 1\frac{1}{3}$

7) $2\frac{10}{11} - 1\frac{1}{3}$

8) $2\frac{1}{2} - 1\frac{1}{2}$

9) $6\frac{3}{5} - 2\frac{1}{5}$

10) $7\frac{2}{5} - 1\frac{1}{5}$

11) $2\frac{5}{7} - 1\frac{1}{3}$

12) $2\frac{1}{4} - 1\frac{1}{2}$

Multiplying Mixed Numbers

		Example:
Helpful	1- Convert the mixed numbers to improper fractions.	
	2- Multiply fractions and simplify if necessary.	$2\frac{1}{3} \times 5\frac{3}{7} =$
Hints	$a\frac{c}{b} = a + \frac{c}{b} = \frac{ab+c}{b}$	$\frac{7}{3} \times \frac{38}{7} = \frac{38}{3} = 12\frac{2}{3}$

✎ **Find each product.**

1) $1\frac{2}{3} \times 1\frac{1}{4}$

2) $1\frac{3}{5} \times 1\frac{2}{3}$

3) $1\frac{2}{3} \times 3\frac{2}{7}$

4) $4\frac{1}{8} \times 1\frac{2}{5}$

5) $2\frac{2}{5} \times 3\frac{1}{5}$

6) $1\frac{1}{3} \times 1\frac{2}{3}$

7) $1\frac{5}{8} \times 2\frac{1}{2}$

8) $3\frac{2}{5} \times 2\frac{1}{5}$

9) $2\frac{2}{3} \times 4\frac{1}{4}$

10) $2\frac{3}{5} \times 1\frac{2}{4}$

11) $1\frac{1}{3} \times 1\frac{1}{4}$

12) $3\frac{2}{5} \times 1\frac{1}{5}$

Dividing Mixed Numbers

Helpful	1- Convert the mixed numbers to improper fractions.	**Example:**
Hints	2- Divide fractions and simplify if necessary.	$2\frac{1}{3} \div 5\frac{3}{7} =$
	$a\frac{c}{b} = a + \frac{c}{b} = \frac{ab+c}{b}$	$\frac{7}{3} \div \frac{38}{7} = \frac{7}{3} \times \frac{7}{38} = \frac{49}{108}$

✏️ *Find each quotient.*

1) $2\frac{1}{5} \div 2\frac{1}{2}$

2) $2\frac{3}{5} \div 1\frac{1}{3}$

3) $3\frac{1}{6} \div 4\frac{2}{3}$

4) $1\frac{2}{3} \div 3\frac{1}{3}$

5) $4\frac{1}{8} \div 2\frac{2}{4}$

6) $3\frac{1}{2} \div 2\frac{3}{5}$

7) $3\frac{5}{9} \div 1\frac{2}{5}$

8) $2\frac{2}{7} \div 1\frac{1}{2}$

9) $3\frac{1}{5} \div 1\frac{1}{2}$

10) $4\frac{3}{5} \div 2\frac{1}{3}$

11) $6\frac{1}{6} \div 1\frac{2}{3}$

12) $2\frac{2}{3} \div 1\frac{1}{3}$

Comparing Decimals

Helpful	-	**Decimals:** is a fraction written in a special form. For example, instead of writing $\frac{1}{2}$ you can write 0.5.	**Example:**
Hints	-	**For comparing:** Equal to = Less than < Greater than > Greater than or equal ≥ Less than or equal ≤	2.67 > 0.267

✎ *Write the correct comparison symbol (>, < or =).*

1) 1.25   2.3

2) 0.5   0.23

3) 3.2   3.2

4) 4.58   45.8

5) 2.75   0.275

6) 5.2   5

7) 3.1   0.31

8) 6.33   0.733

9) 8   0.8

10) 4.56   0.456

11) 1.12   1.14

12) 2.77   2.78

13) 6.08   6.11

14) 1.11   0.211

15) 2.6   2.55

16) 1.24   1.25

17) 5.52   0.552

18) 0.33   0.033

19) 14.4   14.4

20) 0.05   0.50

21) 0.59   0.7

22) 0.5   0.05

23) 0.90   0.9

24) 0.27   0.4

Rounding Decimals

Helpful Hints

We can round decimals to a certain accuracy or number of decimal places. This is used to make calculation easier to do and results easier to understand, when exact values are not too important.

First, you'll need to remember your place values:

12.4567

1: tens 2: ones 4: tenths

5: hundredths 6: thousandths 7: ten thousandths

Example:

$\underline{6}.37 = 6$

✎ *Round each decimal number to the nearest place indicated.*

1) 0.2<u>3</u>

2) 4.<u>0</u>4

3) 5.<u>6</u>23

4) 0.<u>2</u>66

5) <u>6</u>.37

6) 0.8<u>8</u>

7) 8.2<u>4</u>

8) <u>7</u>.0760

9) 1.6<u>2</u>9

10) 6.<u>3</u>959

11) <u>1</u>.9

12) <u>5</u>.2167

13) 5.<u>8</u>63

14) 8.<u>5</u>4

15) 8<u>0</u>.69

16) 6<u>5</u>.85

17) 70.<u>7</u>8

18) 615.7<u>5</u>5

19) 1<u>6</u>.4

20) 9<u>5</u>.81

21) <u>2</u>.408

22) 7<u>6</u>.3

23) 116.<u>5</u>14

24) 8.<u>0</u>6

Adding and Subtracting Decimals

Helpful *Hints*	1– Line up the numbers. 2– Add zeros to have same number of digits for both numbers. 3– Add or Subtract using column addition or subtraction.	**Example:** $\begin{array}{r} 16.18 \\ -\ 13.45 \\ \hline 2.73 \end{array}$

✍ *Add and subtract decimals.*

1) $\begin{array}{r} 15.14 \\ -\ 12.18 \\ \hline \end{array}$

3) $\begin{array}{r} 82.56 \\ +\ 12.28 \\ \hline \end{array}$

5) $\begin{array}{r} 90.37 \\ +\ 56.97 \\ \hline \end{array}$

2) $\begin{array}{r} 65.72 \\ +\ 43.67 \\ \hline \end{array}$

4) $\begin{array}{r} 34.18 \\ -\ 23.45 \\ \hline \end{array}$

6) $\begin{array}{r} 45.78 \\ -\ 23.39 \\ \hline \end{array}$

✍ *Solve.*

7) _____ + 1.3 = 4.8

8) 4.2 + _____ = 11.6

9) 9.9 + _____ = 16

10) 6.9 + _____ = 16.4

11) _____ + 5.1 = 8.6

12) _____ + 7.9 = 15.2

Multiplying and Dividing Decimals

Helpful *Hints*	**For Multiplication:**
	– Set up and multiply the numbers as you do with whole numbers.
	– Count the total number of decimal places in both of the factors.
	– Place the decimal point in the product.
	For Division:
	– If the divisor is not a whole number, move decimal point to right to make it a whole number. Do the same for dividend.
	– Divide similar to whole numbers.

✍ Find each product.

1)
$$\begin{array}{r} 4.5 \\ \times\ 1.6 \\ \hline \end{array}$$

4)
$$\begin{array}{r} 8.9 \\ \times\ 9.7 \\ \hline \end{array}$$

7)
$$\begin{array}{r} 5.7 \\ \times\ 7.8 \\ \hline \end{array}$$

2)
$$\begin{array}{r} 7.7 \\ \times\ 9.9 \\ \hline \end{array}$$

5)
$$\begin{array}{r} 15.1 \\ \times\ 12.6 \\ \hline \end{array}$$

8)
$$\begin{array}{r} 98.20 \\ \times\ 100 \\ \hline \end{array}$$

3)
$$\begin{array}{r} 2.6 \\ \times\ 1.5 \\ \hline \end{array}$$

6)
$$\begin{array}{r} 6.9 \\ \times\ 3.3 \\ \hline \end{array}$$

9)
$$\begin{array}{r} 23.99 \\ \times\ 1000 \\ \hline \end{array}$$

✍ Find each quotient.

10) $9.2 \div 3.6$

11) $27.6 \div 3.8$

12) $12.6 \div 4.7$

13) $6.5 \div 8.1$

14) $1.\,4 \div 10$

15) $3.6 \div 100$

16) $4.24 \div 10$

17) $14.6 \div 100$

18) $1.8 \div 1000$

Converting Between Fractions, Decimals and Mixed Numbers

Helpful *Hints*	**Fraction to Decimal:** − Divide the top number by the bottom number. **Decimal to Fraction:** − Write decimal over 1. − Multiply both top and bottom by 10 for every digit on the right side of the decimal point. − Simplify.

✍ *Convert fractions to decimals.*

1) $\dfrac{9}{10}$

2) $\dfrac{56}{100}$

3) $\dfrac{3}{4}$

4) $\dfrac{2}{5}$

5) $\dfrac{3}{9}$

6) $\dfrac{40}{50}$

7) $\dfrac{12}{10}$

8) $\dfrac{8}{5}$

9) $\dfrac{69}{10}$

✍ *Convert decimal into fraction or mixed numbers.*

10) 0.3

11) 4.5

12) 2.5

13) 2.3

14) 0.8

15) 0.25

16) 0.14

17) 0.2

18) 0.08

19) 0.45

20) 2.6

21) 5.2

Factoring Numbers

Helpful *Hints*	- Factoring numbers means to break the numbers into their prime factors. - First few prime numbers: 2, 3, 5, 7, 11, 13, 17, 19	**Example:** $12 = 2 \times 2 \times 3$

✎ **List all positive factors of each number.**

1) 68

2) 56

3) 24

4) 40

5) 86

6) 78

7) 50

8) 98

9) 45

10) 26

11) 54

12) 28

13) 55

14) 85

15) 48

✎ **List the prime factorization for each number.**

16) 50

17) 25

18) 69

19) 21

20) 45

21) 68

22) 26

23) 86

24) 93

Greatest Common Factor

Helpful	- List the prime factors of each number.	**Example:**
	- Multiply common prime factors.	
Hints		$200 = 2 \times 2 \times 2 \times 5 \times 5$
		$60 = 2 \times 2 \times 3 \times 5$
		GCF $(200, 60) = 2 \times 2 \times 5 = 20$

✍ *Find the GCF for each number pair.*

1) 20, 30

2) 4, 14

3) 5, 45

4) 68, 12

5) 5, 12

6) 15, 27

7) 3, 24

8) 34, 6

9) 4, 10

10) 5, 3

11) 6, 16

12) 30, 3

13) 24, 28

14) 70, 10

15) 45, 8

16) 90, 35

17) 78, 34

18) 55, 75

19) 60, 72

20) 100, 78

21) 30, 40

Least Common Multiple

Helpful *Hints*	- Find the GCF for the two numbers. - Divide that GCF into either number. - Take that answer and multiply it by the other number.	**Example:** LCM (200, 60): GCF is 20 200 ÷ 20 = 10 10 × 60 = 600

✏️ *Find the LCM for each number pair.*

1) 4, 14

2) 5, 15

3) 16, 10

4) 4, 34

5) 8, 3

6) 12, 24

7) 9, 18

8) 5, 6

9) 8, 19

10) 9, 21

11) 19, 29

12) 7, 6

13) 25, 6

14) 4, 8

15) 30, 10, 50

16) 18, 36, 27

17) 12, 8, 18

18) 8, 18, 4

19) 26, 20, 30

20) 10, 4, 24

21) 15, 30, 45

Divisibility Rules

Helpful	- Divisibility means that a number can be divided by other numbers evenly.	**Example:**
Hints		24 is divisible by 6, because 24 ÷ 6 = 4

✏️ *Use the divisibility rules to find the factors of each number.*

8	<u>2</u> 3 <u>4</u> 5 6 7 <u>8</u> 9 10
1) 16	2 3 4 5 6 7 8 9 10
2) 10	2 3 4 5 6 7 8 9 10
3) 15	2 3 4 5 6 7 8 9 10
4) 28	2 3 4 5 6 7 8 9 10
5) 36	2 3 4 5 6 7 8 9 10
6) 15	2 3 4 5 6 7 8 9 10
7) 27	2 3 4 5 6 7 8 9 10
8) 70	2 3 4 5 6 7 8 9 10
9) 57	2 3 4 5 6 7 8 9 10
10) 102	2 3 4 5 6 7 8 9 10
11) 144	2 3 4 5 6 7 8 9 10
12) 75	2 3 4 5 6 7 8 9 10

Test Preparation

1) Mr. Jones saves $2,500 out of his monthly family income of $55,000. What fractional part of his income does he save?

A. $\frac{1}{22}$

B. $\frac{1}{11}$

C. $\frac{3}{25}$

D. $\frac{2}{15}$

2) What is the missing prime factor of number 360?

$360 = 2^3 \times 3^2 \times$ ___5___

Write your answer in the box below.

$$\boxed{5}$$

3) Which list shows the fractions in order from least to greatest?

$$\frac{2}{3}, \quad \frac{5}{7}, \quad \frac{3}{10}, \quad \frac{1}{2}, \quad \frac{6}{13}$$

A. $\quad \frac{2}{3}, \quad \frac{5}{7}, \quad \frac{3}{10}, \quad \frac{1}{2}, \quad \frac{6}{13}$

B. $\quad \frac{6}{13}, \quad \frac{1}{2}, \quad \frac{2}{3}, \quad \frac{5}{7}, \quad \frac{3}{10}$

C. $\quad \frac{3}{10}, \quad \frac{2}{3}, \quad \frac{5}{7}, \quad \frac{1}{2}, \quad \frac{6}{13}$

D. $\quad \frac{3}{10}, \quad \frac{6}{13}, \quad \frac{1}{2}, \quad \frac{2}{3}, \quad \frac{5}{7}$

4) Which statement about 5 multiplied by $\frac{2}{3}$ is true?

A. The product is between 2 and 3.

B. The product is between 3 and 4

C. The product is more than $\frac{11}{3}$.

D. The product is between $\frac{14}{3}$ and 5.

5) Four one – foot rulers can be split among how many users to leave each with $\frac{1}{6}$ of a ruler?

   A. 4

   B. 6

   C. 12

   D. 24

6) Last week 24,000 fans attended a football match. This week three times as many bought tickets, but one sixth of them cancelled their tickets. How many are attending this week?

   A. 48,000

   B. 54,000

   C. 60,000

   D. 72,000

7) What is the missing prime factor of number 180?

     $180 = 2^2 \times 3^2 \times$ ___

   A. 2

   B. 3

   C. 5

   D. 6

Answers of Worksheets – Chapter 1

Simplifying Fractions

1) $\frac{11}{18}$

2) $\frac{4}{5}$

3) $\frac{2}{3}$

4) $\frac{3}{4}$

5) $\frac{1}{3}$

6) $\frac{1}{4}$

7) $\frac{4}{9}$

8) $\frac{1}{2}$

9) $\frac{2}{5}$

10) $\frac{1}{9}$

11) $\frac{5}{9}$

12) $\frac{3}{4}$

13) $\frac{5}{8}$

14) $\frac{13}{16}$

15) $\frac{1}{5}$

16) $\frac{4}{7}$

17) $\frac{1}{2}$

18) $\frac{5}{12}$

19) $\frac{3}{8}$

20) $\frac{1}{4}$

21) $\frac{5}{9}$

Adding and Subtracting Fractions

1) $\frac{7}{6}$

2) $\frac{14}{15}$

3) $\frac{4}{3}$

4) $\frac{83}{36}$

5) $\frac{3}{5}$

6) $\frac{13}{14}$

7) $\frac{23}{20}$

8) $\frac{13}{15}$

9) $\frac{31}{25}$

10) $\frac{2}{5}$

11) $\frac{11}{35}$

12) $\frac{1}{6}$

13) $\frac{13}{45}$

14) $\frac{3}{14}$

15) $\frac{1}{6}$

16) $\frac{1}{36}$

17) $\frac{9}{40}$

18) $\frac{7}{18}$

Multiplying and Dividing Fractions

1) $\frac{2}{15}$

2) $\frac{1}{2}$

3) $\frac{6}{35}$

4) $\frac{1}{8}$

5) $\frac{6}{25}$

6) $\frac{7}{27}$

7) $\frac{1}{4}$

8) $\frac{1}{12}$

9) $\frac{5}{12}$

10) $\frac{8}{9}$

11) $\frac{3}{2}$

12) $\frac{8}{11}$

13) $\frac{55}{7}$

14) $\frac{27}{25}$

15) 1

16) 3

17) $\frac{4}{3}$

18) $\frac{25}{63}$

Adding Mixed Numbers

1) 10

2) $5\frac{1}{2}$

3) $9\frac{3}{5}$

4) 4

5) $10\frac{2}{3}$

6) $4\frac{2}{3}$

7) $3\frac{8}{33}$

8) 4

9) $10\frac{4}{5}$

10) $7\frac{1}{5}$

11) $2\frac{1}{21}$

12) $3\frac{3}{4}$

Subtract Mixed Numbers

1) 1

2) $\frac{1}{4}$

3) $1\frac{2}{5}$

4) $\frac{2}{3}$

5) $\frac{2}{3}$

6) 2

7) $1\frac{19}{33}$

8) 1

9) $4\frac{2}{5}$

10) $6\frac{1}{5}$

11) $1\frac{8}{21}$

12) $\frac{3}{4}$

Multiplying Mixed Numbers

1) $2\frac{1}{12}$

2) $2\frac{2}{3}$

3) $5\frac{10}{21}$

4) $5\frac{31}{40}$

5) $7\frac{17}{25}$

6) $2\frac{2}{9}$

7) $4\frac{1}{16}$

8) $7\frac{12}{25}$

9) $11\frac{1}{3}$

10) $3\frac{9}{10}$

11) $1\frac{2}{3}$

12) $4\frac{2}{25}$

Dividing Mixed Numbers

1) $\frac{22}{25}$

2) $1\frac{19}{20}$

3) $\frac{19}{28}$

4) $\frac{1}{2}$

5) $1\frac{13}{20}$

6) $1\frac{9}{26}$

7) $2\frac{34}{63}$

8) $1\frac{11}{21}$

9) $2\frac{2}{15}$

10) $1\frac{34}{35}$

11) $3\frac{7}{10}$

12) 2

Comparing Decimals

1) 1.25 < 2.3

2) 0.5 > 0.23

3) 3.2 = 3.2

4) 4.58 < 45.8

5) 2.75 > 0.275

6) 5.2 > 5

7) 3.1 > 0.31

8) 6.33 > 0.733

9) 8 > 0.8

10) 4.56 > 0.456

11) 1.12 < 1.14

12) 2.77 < 2.78

13) 6.08 < 6.11

14) 1.11 > 0.211

15) 2.6 > 2.55

16) 1.24 < 1.25

17) 5.52 > 0.552

18) 0.33 > 0.033

19) 14.4 = 14.4

20) 0.05 < 0.50

21) 0.59 < 0.7

22) 0.5 > 0.05

23) 0.90 = 0.9

24) 0.27 < 0.4

Rounding Decimals

1) 0.2
2) 4.0
3) 5.6
4) 0.3
5) 6
6) 0.9
7) 8.2
8) 7

9) 1.63
10) 6.4
11) 2
12) 5
13) 5.9
14) 8.5
15) 81
16) 66

17) 70.8
18) 616
19) 16
20) 96
21) 2
22) 76
23) 116.5
24) 8.1

Adding and Subtracting Decimals

1) 2.96
2) 109.39
3) 94.84
4) 10.73

5) 147.34
6) 22.39
7) 3.5
8) 7.4

9) 6.1
10) 9.5
11) 3.5
12) 7.3

Multiplying and Dividing Decimals

1) 7.2
2) 76.23
3) 3.9
4) 86.33
5) 190.26
6) 22.77

7) 44.46
8) 9820
9) 23990
10) 2.5555...
11) 7.2631...
12) 2.6808...

13) 0.8024...
14) 0.14
15) 0.036
16) 0.424
17) 0.146
18) 0.0018

Converting Between Fractions, Decimals and Mixed Numbers

1) 0.9
2) 0.56
3) 0.75
4) 0.4
5) 0.333...
6) 0.8

7) 1.2
8) 1.6
9) 6.9
10) $\frac{3}{10}$
11) $4\frac{1}{2}$

12) $2\frac{1}{2}$
13) $2\frac{3}{10}$
14) $\frac{4}{5}$
15) $\frac{1}{4}$

16) $\dfrac{7}{50}$

17) $\dfrac{1}{5}$

18) $\dfrac{2}{25}$

19) $\dfrac{9}{20}$

20) $2\dfrac{3}{5}$

21) $5\dfrac{1}{5}$

Factoring Numbers

1) 1, 2, 4, 17, 34, 68
2) 1, 2, 4, 7, 8, 14, 28, 56
3) 1, 2, 3, 4, 6, 8, 12, 24
4) 1, 2, 4, 5, 8, 10, 20, 40
5) 1, 2, 43, 86
6) 1, 2, 3, 6, 13, 26, 39, 78
7) 1, 2, 5, 10, 25, 50
8) 1, 2, 7, 14, 49, 98
9) 1, 3, 5, 9, 15, 45
10) 1, 2, 13, 26
11) 1, 2, 3, 6, 9, 18, 27, 54
12) 1, 2, 4, 7, 14, 28

13) 1, 5, 11, 55
14) 1, 5, 17, 85
15) 1, 2, 3, 4, 6, 8, 12, 16, 24, 48
16) $2 \times 5 \times 5$
17) 5×5
18) 3×23
19) 3×7
20) $3 \times 3 \times 5$
21) $2 \times 2 \times 17$
22) 2×13
23) 2×43
24) 3×31

Greatest Common Factor

1) 10
2) 2
3) 5
4) 4
5) 1
6) 3
7) 3

8) 2
9) 2
10) 1
11) 2
12) 3
13) 4
14) 10

15) 1
16) 5
17) 2
18) 5
19) 12
20) 2
21) 10

Least Common Multiple

1) 28
2) 15
3) 80
4) 68
5) 24
6) 24
7) 18

8) 30
9) 152
10) 63
11) 551
12) 42
13) 150
14) 8

15) 150
16) 108
17) 72
18) 72
19) 780
20) 120
21) 90

Divisibility Rules

1) 16   <u>2</u> 3 <u>4</u> 5 6 7 <u>8</u> 9 10

2) 10   <u>2</u> 3 4 <u>5</u> 6 7 8 9 <u>10</u>

3) 15   2 <u>3</u> 4 <u>5</u> 6 7 8 9 10

4) 28   <u>2</u> 3 <u>4</u> 5 6 <u>7</u> 8 9 10

5) 36   <u>2</u> <u>3</u> <u>4</u> 5 <u>6</u> 7 8 <u>9</u> 10

6) 18   <u>2</u> <u>3</u> 4 5 <u>6</u> 7 8 <u>9</u> 10

7) 27   2 <u>3</u> 4 5 6 7 8 <u>9</u> 10

8) 70   <u>2</u> 3 4 <u>5</u> 6 <u>7</u> 8 9 <u>10</u>

9) 57   2 <u>3</u> 4 5 6 7 8 9 10

10) 102   <u>2</u> <u>3</u> 4 5 <u>6</u> 7 8 9 10

11) 144   <u>2</u> <u>3</u> <u>4</u> 5 <u>6</u> 7 <u>8</u> <u>9</u> 10

12) 75   2 <u>3</u> 4 <u>5</u> 6 7 8 9 10

Test Preparation Answers

1) Choice A is correct

2,500 out of 55,000 equals to $\dfrac{2500}{55000} = \dfrac{25}{550} = \dfrac{1}{22}$

2) The answer is 5^1.

Let x be the number of blank.

$360 = 2 \times 2 \times 2 \times 3 \times 3 \times x \Rightarrow x = \dfrac{360}{72} \Rightarrow x = 5$

3) Choice D is correct.

Compare each fraction, then we have:

$\dfrac{3}{10} < \dfrac{6}{13} < \dfrac{1}{2} < \dfrac{2}{3} < \dfrac{5}{7}$

4) Choice B is correct

To find the discount, multiply the 5 by $\dfrac{2}{3}$. Therefore we have $\dfrac{10}{3}$

After simplification now we have $3\dfrac{1}{3}$,that is between 3 and 4.

5) Choice D is correct

$4 \div \dfrac{1}{6} = 24$

6) Choice C is correct

Three times of 24,000 is 72,000. One sixth of them cancelled their tickets.

One sixth of 72,000 equals 12,000 (1/6 × 72,000 = 12,000).

60,000 (72000 − 12000 = 60000) fans are attending this week

7) Choice C is correct

Let x be the missed number. $180 = 4 \times 9 \times x \Rightarrow x=5$

Chapter 2: Integers and Absolute Value

Topics that you'll learn in this chapter:

- ✓ Adding and Subtracting Integers
- ✓ Multiplying and Dividing Integers
- ✓ Ordering Integers and Numbers
- ✓ Arrange, Order, and Comparing Integers
- ✓ Order of Operations
- ✓ Mixed Integer Computations
- ✓ Absolute Value
- ✓ Integers and Absolute Value
- ✓ Classifying Real Numbers Venn Diagram

Adding and Subtracting Integers

Helpful	-	**Integers:** {... , −3, −2, −1, 0, 1, 2, 3, ...} Includes: zero, counting numbers, and the negative of the counting numbers.	**Example:**
Hints			$12 + 10 = 22$
		− Add a positive integer by moving to the right on the number line.	$25 - 13 = 12$
		− Add a negative integer by moving to the left on the number line.	$(-24) + 12 = -12$
			$(-14) + (-12) = -26$
		− Subtract an integer by adding its opposite.	$14 - (-13) = 27$

✍ *Find the sum.*

1) $(-12) + (-4)$

2) $5 + (-24)$

3) $(-14) + 23$

4) $(-8) + (39)$

5) $43 + (-12)$

6) $(-23) + (-4) + 3$

7) $4 + (-12) + (-10) + (-25)$

8) $19 + (-15) + 25 + 11$

9) $(-9) + (-12) + (32 - 14)$

10) $4 + (-30) + (45 - 34)$

✍ *Find the difference.*

11) $(-14) - (-9) - (18)$

12) $(-9) - (-25)$

13) $(-12) - (8)$

14) $(28) - (-4)$

15) $(34) - (2)$

16) $(55) - (-5) + (-4)$

17) $(9) - (2) - (-5)$

18) $(2) - (4) - (-15)$

19) $(23) - (4) - (-34)$

20) $(-45) - (-87)$

Multiplying and Dividing Integers

Helpful	(negative) × (negative) = positive	**Examples:**
	(negative) ÷ (negative) = positive	$3 \times 2 = 6$
Hints	(negative) × (positive) = negative	$3 \times -3 = -9$
	(negative) ÷ (positive) = negative	$-2 \times -2 = 4$
	(positive) × (positive) = positive	$10 \div 2 = 5$
		$-4 \div 2 = -2$
		$-12 \div -6 = 3$

✍️ *Find each product.*

1) $(-8) \times (-2)$

2) 3×6

3) $(-4) \times 5 \times (-6)$

4) $2 \times (-6) \times (-6)$

5) $11 \times (-12)$

6) $10 \times (-5)$

7) 8×8

8) $(-8) \times (-9)$

9) $6 \times (-5) \times 3$

10) $6 \times (-1) \times 2$

✍️ *Find each quotient.*

11) $18 \div 3$

12) $(-24) \div 4$

13) $(-63) \div (-9)$

14) $54 \div 9$

15) $20 \div (-2)$

16) $(-66) \div (-11)$

17) $64 \div 8$

18) $(-121) \div 11$

19) $72 \div 9$

20) $16 \div 4$

Ordering Integers and Numbers

Helpful *Hints*	To compare numbers, you can use number line! As you move from left to right on the number line, you find a bigger number!	**Example:** Order integers from least to greatest. $(-11, -13, 7, -2, 12)$ $-13 < -11 < -2 < 7 < 12$

✏️ **Order each set of integers from least to greatest.**

1) $-15, -19, 20, -4, 1$ ___, ___, ___, ___, ___, ___

2) $6, -5, 4, -3, 2$ ___, ___, ___, ___, ___, ___

3) $15, -42, 19, 0, -22$ ___, ___, ___, ___, ___, ___

4) $26, -91, 0, -13, 67, -55$ ___, ___, ___, ___, ___, ___

5) $-17, -71, 90, -25, -54, -39$ ___, ___, ___, ___, ___, ___

6) $98, 5, 46, 19, 77, 24$ ___, ___, ___, ___, ___, ___

✏️ **Order each set of integers from greatest to least.**

7) $-2, 5, -3, 6, -4$ ___, ___, ___, ___, ___, ___

8) $-37, 7, -17, 27, 47$ ___, ___, ___, ___, ___, ___

9) $32, -27, 19, -17, 15$ ___, ___, ___, ___, ___, ___

10) $68, 81, 21, -18, 94, 72$ ___, ___, ___, ___, ___, ___

Arrange, Order, and Comparing Integers

Helpful *Hints*	When using a number line, numbers increase as you move to the right.	**Examples:** $5 < 7,$ $-5 < -2$ $-18 < -12$

🖎*Arrange these integers in descending order.*

1) $21, 71, -18, -10, 82$ ___, ___, ___, ___, ___, ___

2) $15, 11, 20, 12, -9, -5$ ___, ___, ___, ___, ___, ___

3) $-5, 20, 15, 9, -11$ ___, ___, ___, ___, ___, ___

4) $19, 18, -9, -6, -11$ ___, ___, ___, ___, ___, ___

5) $56, -34, -12, -5, 32$ ___, ___, ___, ___, ___, ___

🖎*Compare. Use >, =, <*

6) -8 ____ 12 11) -56 ____ -58

7) -10 ____ -16 12) 78 ____ 87

8) 43 ____ 34 13) -92 ____ -102

9) 15 ____ -16 14) -12 ____ -12

10) -354 ____ -345 15) -721 ____ -821

Order of Operations

Helpful	-	Use "order of operations" rule when there are more than one math operation.	**Example:**
Hints	-	PEMDAS (parentheses / exponents / multiply / divide / add / subtract)	$(12 + 4) \div (- 4) = - 4$

✎*Evaluate each expression.*

1) $(2 \times 2) + 5$

2) $24 - (3 \times 3)$

3) $(6 \times 4) + 8$

4) $25 - (4 \times 2)$

5) $(6 \times 5) + 3$

6) $64 - (2 \times 4)$

7) $25 + (1 \times 8)$

8) $(6 \times 7) + 7$

9) $48 \div (4 + 4)$

10) $(7 + 11) \div (- 2)$

11) $9 + (2 \times 5) + 10$

12) $(5 + 8) \times \frac{3}{5} + 2$

13) $2 \times 7 - (\frac{10}{9 - 4})$

14) $(12 + 2 - 5) \times 7 - 1$

15) $(\frac{7}{5 - 1}) \times (2 + 6) \times 2$

16) $20 \div (4 - (10 - 8))$

17) $\frac{50}{4\,(5 - 4) - 3}$

18) $2 + (8 \times 2)$

Mixed Integer Computations

Helpful *Hints*	**It worth remembering:**	**Example:**
	(negative) × (negative) = positive	
	(negative) ÷ (negative) = positive	$(-5) + 6 = 1$
	(negative) × (positive) = negative	$(-3) \times (-2) = 6$
	(negative) ÷ (positive) = negative	$(9) \div (-3) = -3$
	(positive) × (positive) = positive	

✐ *Compute.*

1) $(-70) \div (-5)$

2) $(-14) \times 3$

3) $(-4) \times (-15)$

4) $(-65) \div 5$

5) $18 \times (-7)$

6) $(-12) \times (-2)$

7) $\dfrac{(-60)}{(-20)}$

8) $24 \div (-8)$

9) $22 \div (-11)$

10) $\dfrac{(-27)}{3}$

11) $4 \times (-4)$

12) $\dfrac{(-48)}{12}$

13) $(-14) \times (-2)$

14) $(-7) \times (7)$

15) $\dfrac{-30}{-6}$

16) $(-54) \div 6$

17) $(-60) \div (-5)$

18) $(-7) \times (-12)$

19) $(-14) \times 5$

20) $88 \div (-8)$

Absolute Value

Helpful *Hints*	Refers to the distance of a number from 0, the distances are positive. Therefore, absolute value of a number cannot be negative. $\|-22\| = 22$	**Example:** $\|12\| \times \|-2\| = 24$

$$|x| = \begin{cases} x & for\ x \geq 0 \\ -x & for\ x < 0 \end{cases}$$

$$|x| < n \quad \Rightarrow -n < x < n$$

$$|x| > n \quad \Rightarrow x < -n\ or\ x > n$$

✎*Evaluate.*

1) $|-4| + |-12| - 7$

2) $|-5| + |-13|$

3) $-18 + |-5 + 3| - 8$

4) $|27| \div |9|$

5) $|-9| \div |-1|$

6) $|200| \div |-100|$

7) $|55| \div |11|$

8) $|36| \div |-6|$

9) $|25| \times |-5|$

10) $|-3| \times |-8|$

11) $|12| \times |-5|$

12) $|11| \times |-6|$

13) $|-8| \times |4|$

14) $|-9| \times |-7|$

15) $|43 - 67 + 9| + |-11| - 1$

16) $|-45 + 78| + |23| - |45|$

17) $75 + |-11 - 30| - |2|$

18) $|-3 + 15| + |9 + 4| - 1$

Integers and Absolute Value

Helpful	To find an absolute value of a number, just find it's distance from 0!	Example:
Hints		$\lvert-6\rvert = 6$
		$\lvert 6\rvert = 6$
		$\lvert-12\rvert = 12$
		$\lvert 12\rvert = 12$

✎ *Write absolute value of each number.*

1) -4

2) -7

3) -8

4) 4

5) 5

6) -10

7) 1

8) 6

9) 8

10) -2

11) -1

12) 10

13) 3

14) 7

15) -5

16) -3

17) -9

18) 2

19) 4

20) -6

21) 9

✎ *Evaluate.*

22) $\lvert-43\rvert - \lvert 12\rvert + 10$

23) $76 + \lvert-15 - 45\rvert - \lvert 3\rvert$

24) $30 + \lvert-62\rvert - 46$

25) $\lvert 32\rvert - \lvert-78\rvert + 90$

26) $\lvert-35 + 4\rvert + 6 - 4$

27) $\lvert-4\rvert + \lvert-11\rvert$

28) $\lvert-6 + 3 - 4\rvert + \lvert 7 + 7\rvert$

29) $\lvert-9\rvert + \lvert-19\rvert - 5$

Classifying Real Numbers Venn Diagram

Helpful

Hints

Example:

0.25 =

rational

number

and

real number

Natural numbers (counting numbers): are the numbers that are used for counting. 1, 2, 3, …, 100, … are natural numbers.

Whole numbers are the natural numbers plus zero.

Integers include all whole numbers plus "negatives" of the natural numbers.

Rational numbers are numbers that can be written as a fraction. Both top and bottom numbers must be integers.

Irrational numbers are all numbers which cannot be written as fractions.

Real numbers include both the rational and irrational numbers.

✍️*Identify all of the subsets of real number system to which each number belongs.*

Example:

0.1259 : Rational number

$\sqrt{2}$: Irrational number

3 : Natural number, whole number, Integer, rational number

1) 0

2) − 5

3) − 8.5

4) $\sqrt{4}$

5) − 10

6) 18

7) 6

8) π

9) $1\frac{2}{7}$

10) − 1

11) $\sqrt{5}$

Test Preparation

1) Which expression has a value of (− 18)?

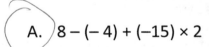

    A. $8 - (-4) + (-15) \times 2$

    B. $12 + (-3) \times (-2)$

    C. $-6 \times (-6) \times (-2) \div (-4)$

    D. $(-2) \times (-7) + 4$

2) $4 + 8 \times (3) - [8 + 8 \times 5] \div 6 = ?$

Write your answer in the box below.

$$\boxed{20}$$

Answers of Worksheets – Chapter 2

Adding and Subtracting Integers

1) -16

2) -19

3) 9

4) 31

5) 31

6) -24

7) -43

8) 40

9) -3

10) -15

11) -23

12) 16

13) -20

14) 32

15) 32

16) 56

17) 12

18) 13

19) 53

20) 42

Multiplying and Dividing Integers

1) 16

2) 18

3) 120

4) 72

5) -132

6) -50

7) 64

8) 72

9) -90

10) -12

11) 6

12) -6

13) 7

14) 6

15) -10

16) 6

17) 8

18) -11

19) 8

20) 4

Ordering Integers and Numbers

1) $-19, -15, -4, 1, 20$

2) $-5, -3, 2, 4, 6$

3) $-42, -22, 0, 15, 19$

4) $-91, -55, -13, 0, 26, 67$

5) $-71, -54, -39, -25, -17, 90$

6) $5, 19, 24, 46, 77, 98$

7) $6, 5, -2, -3, -4$

8) $47, 27, 7, -17, -37$

9) $32, 19, 15, -17, -27$

10) $94, 81, 72, 68, 21, -18$

Arrange and Order, Comparing Integers

1) 82, 71, 21, − 10, − 18

2) 20, 15, 12, 11, − 5, − 9

3) 20, 15, 9, − 5, −11

4) 19, 18, − 6, − 9, − 11

5) 56, 32, − 5, − 12, − 34

6) <

7) >

8) >

9) >

10) <

11) >

12) <

13) >

14) =

15) >

Order of Operations

1) 9

2) 15

3) 32

4) 17

5) 33

6) 56

7) 33

8) 49

9) 6

10) − 9

11) 29

12) 9.8

13) 12

14) 62

15) 28

16) 10

17) 50

18) 18

Mixed Integer Computations

1) 14

2) − 42

3) 60

4) − 13

5) − 126

6) 24

7) 3

8) − 3

9) − 2

10) − 9

11) − 16

12) − 4

13) 28

14) − 49

15) 5

16) − 9

17) 12

18) 84

19) − 70

20) − 11

Absolute Value

1) 9	7) 5	13) 32
2) 18	8) 6	14) 63
3) − 24	9) 125	15) 25
4) 3	10) 24	16) 11
5) 9	11) 60	17) 114
6) 2	12) 66	18) 24

Integers and Absolute Value

1) 4	11) 1	21) 9
2) 7	12) 10	22) 41
3) 8	13) 3	23) 133
4) 4	14) 7	24) 46
5) 5	15) 5	25) 44
6) 10	16) 3	26) 33
7) 1	17) 9	27) 15
8) 6	18) 2	28) 21
9) 8	19) 4	29) 23
10) 2	20) 6	

Classifying Real Numbers Venn Diagram

1) 0: whole number, integer, rational number
2) − 5: integer, rational number
3) − 8.5: rational number
4) $\sqrt{4}$: natural number, whole number, integer, rational number
5) − 10: integer, rational number
6) 18 : natural number, whole number, integer, rational number
7) 6: natural number, whole number, integer, rational number
8) π: irrational number
9) $1\frac{2}{7}$: rational number
10)   − 1: integer, rational number
11) $\sqrt{5}$: irrational number

Test Preparation Answers

1) **Choice A is correct**

Use PEMDAS (order of operation):

$$8 - (-4) + (-15) \times 2 = 8 + 4 - 30 = -18$$

2) **The answer is 20.**

Use PEMDAS (order of operation):

$$4 + 8 \times (3) - [48] \div 6 = 4 + 24 - 8 = 20$$

Chapter 3: Percent

Topics that you'll learn in this chapter:

- ✓ Percentage Calculations
- ✓ Converting Between Percent, Fractions, and Decimals
- ✓ Percent Problems
- ✓ Find What Percentage a Number Is of Another
- ✓ Find a Percentage of a Given Number
- ✓ Percent of Increase and Decrease

Percentage Calculations

Helpful	-	Use the following formula to find part, whole, or percent:	**Example:**
Hints		$$part = \frac{percent}{100} \times whole$$	$$\frac{20}{100} \times 100 = 20$$

✎ *Calculate the percentages.*

1) 50% of 25	7) 65% of 8	13) 20% of 70
2) 80% of 15	8) 78% of 54	14) 55% of 60
3) 30% of 34	9) 50% of 80	15) 80% of 10
4) 70% of 45	10) 20% of 10	16) 20% of 880
5) 10% of 0	11) 40% of 40	17) 70% of 100
6) 80% of 22	12) 90% of 0	18) 80% of 90

✎ *Solve.*

19) 50 is what percentage of 75?

20) What percentage of 100 is 70

21) Find what percentage of 60 is 35.

22) 40 is what percentage of 80?

Converting Between Percent, Fractions, and Decimals

Helpful	– To a percent: Move the decimal point 2 places to the right and add the % symbol.	**Examples:**
Hints	– Divide by 100 to convert a number from percent to decimal.	30% = 0.3
		0.24 = 24%

✍ Converting fractions to decimals.

1) $\dfrac{50}{100}$

2) $\dfrac{38}{100}$

3) $\dfrac{15}{100}$

4) $\dfrac{80}{100}$

5) $\dfrac{7}{100}$

6) $\dfrac{35}{100}$

7) $\dfrac{90}{100}$

8) $\dfrac{20}{100}$

9) $\dfrac{7}{100}$

✍ Write each decimal as a percent.

10) 0.5

11) 0.9

12) 0.002

13) 0.524

14) 0.1

15) 0.03

16) 3.63

17) 0.008

18) 4.78

Percent Problems

Helpful	Base = Part ÷ Percent Part = Percent × Base Percent = Part ÷ Base	**Example:** 2 is 10% of 20. 2 ÷ 0.10 = 20 2 = 0.10 × 20 0.10 = 2 ÷ 20
Hints		

✎*Solve each problem.*

1) 51 is 340% of what?

2) 93% of what number is 97?

3) 27% of 142 is what number?

4) What percent of 125 is 29.3?

5) 60 is what percent of 126?

6) 67 is 67% of what?

7) 67 is 13% of what?

8) 41% of 78 is what?

9) 1 is what percent of 52.6?

10) What is 59% of 14 m?

11) What is 90% of 130 inches?

12) 16 inches is 35% of what?

13) 90% of 54.4 hours is what?

14) What percent of 33.5 is 21?

15) Liam scored 22 out of 30 marks in Algebra, 35 out of 40 marks in science and 89 out of 100 marks in mathematics. In which subject his percentage of marks in best?

16) Ella require 50% to pass. If she gets 280 marks and falls short by 20 marks, what were the maximum marks she could have got?

Find What Percentage a Number Is of Another

Helpful *Hints*	PERCENT: the number with the percent sign (%). PART: the number with the word "is". WHOLE: the number with the word "of". – Divide the Part by the Base. – Convert the answer to percent.	**Example:** 20 is what percent of 50? $20 \div 50 = 0.40 = 40\%$

Find the percentage of the numbers.

1) 5 is what percent of 90?

2) 15 is what percent of 75?

3) 20 is what percent of 400?

4) 18 is what percent of 90?

5) 3 is what percent of 15?

6) 8 is what percent of 80?

7) 11 is what percent of 55?

8) 9 is what percent of 90?

9) 2.5 is what percent of 10?

10) 5 is what percent of 25?

11) 60 is what percent of 20?

12) 12 is what percent of 48?

13) 14 is what percent of 28?

14) 8.2 is what percent of 32.8?

15) 1200 is what percent of 4,800?

16) 4,000 is what percent of 20,000?

17) 45 is what percent of 900?

18) 10 is what percent of 200?

19) 15 is what percent of 60?

20) 1.2 is what percent of 24?

Find a Percentage of a Given Number

Helpful	-	Use following formula to find part, whole, or percent: $$\text{part} = \frac{\text{percent}}{100} \times \text{whole}$$	**Example:** $$\frac{50}{100} \times 50 = 25$$
Hints			

Find a Percentage of a Given Number.

1) 90% of 50

2) 40% of 50

3) 10% of 0

4) 80% of 80

5) 60% of 40

6) 50% of 60

7) 30% of 20

8) 35% of 10

9) 10% of 80

10) 10% of 60

11) 100% 0f 50

12) 90% of 34

13) 80% of 42

14) 90% of 12

15) 20% of 56

16) 40% of 40

17) 40% of 6

18) 70% of 38

19) 30% of 3

20) 40% of 50

21) 100% of 8

Percent of Increase and Decrease

Helpful

Hints

– To find the percentage increase:

New Number – Original Number

The result ÷ Original Number × 100

If your answer is a negative number, then this is a percentage decrease.

To calculate percentage decrease:

Original Number – New Number

The result ÷ Original Number × 100

Example:

From 84 miles to 24 miles = 71.43% decrease

Find each percent change to the nearest percent. Increase or decrease.

1) From 32 grams to 82 grams.

2) From 150 m to 45 m

3) From $438 to $443

4) From 256 ft to 140 ft

5) From 6469 ft to 7488 ft

6) From 36 inches to 90 inches

7) From 54 ft to 104 ft

8) From 84 miles to 24 miles

9) The population of a place in a particular year increased by 15%. Next year it decreased by 15%. Find the net increase or decrease percent in the initial population.

10)   The salary of a doctor is increased by 40%. By what percent should the new salary be reduced in order to restore the original salary?

Test Preparation

1) Jason needs an 75% average in his writing class to pass. On his first 4 exams, he earned scores of 68%, 72%, 85%, and 90%. What is the minimum score Jason can earn on his fifth and final test to pass?

   Write your answer in the box below.

   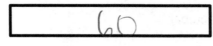

   > 60

2) A shirt costing $200 is discounted 15%. Which of the following expressions can be used to find the selling price of the shirt?

   A. $(200)(0.70)$
   B. $(200) - 200(0.30)$
   C. $(200)(0.15) - (200)(0.15)$
   D. $(200)(0.85)$

3) The price of a car was $20,000 in 2014 and $16,000 in 2015. What is the rate of depreciation of the price of the car per year?

   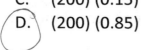

   A. 15 %
   B. 20 %
   C. 25 %
   D. 30 %

4) The price of a laptop is decreased by 10% to $360. What is its original price?

$\frac{40}{360}$

A. 320

B. 380

C. 400

D. 450

5) A bank is offering 4.5% simple interest on a savings account. If you deposit $8,000, how much interest will you earn in five years?

A. $360

B. $720

C. $1,800

D. $3,600

$$8000$$
$$.045$$
$$450000$$
$$32000\ 00$$
$$3600.00$$

6) There are 60 boys and 90 girls in a class. Of these students, 18 boys and 12 girls write left–handed. What percentage of the students in this class write left–handed?
Write your answer in the box below.

$\frac{18}{12}$
$\overline{30}$

150

20%

7) 25 is What percent of 20?

    A. 20 %

    B. 25 %

    C. 125 %

    D. 150 %

8) A $40 shirt now selling for $28 is discounted by what percent?

    A. 20 %

    B. 30 %

    C. 40 %

    D. 60 %

$$\frac{28}{40} = \frac{7}{10}$$

9) From last year, the price of gasoline has increased from $1.25 per gallon to $1.75 per gallon. The new price is what percent of the original price?

    A. 72 %

    B. 120 %

    C. 140 %

    D. 160 %

$$\frac{1.25}{1.75} = \frac{25}{35} = \frac{5}{7}$$

10) Sophia purchased a sofa for $530.40. The sofa is regularly priced at $624. What was the percent discount Sophia received on the sofa?

A. 12%

B. 15%

C. 20%

D. 25%

11) 55 students took an exam and 11 of them failed. What percent of the students passed the exam?

A. 20 %

B. 40 %

C. 60 %

D. 80 %

12) The price of a sofa is decreased by 25% to $420. What was its original price?

A. $480

B. $520

C. $560

D. $600

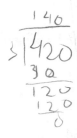

Answers of Worksheets – Chapter 3

Percentage Calculations

1) 12.5	9) 40	17) 70
2) 12	10) 2	18) 72
3) 10.2	11) 16	19) 67%
4) 31.5	12) 0	20) 70%
5) 0	13) 14	21) 58%
6) 17.6	14) 33	22) 50%
7) 5.2	15) 8	
8) 42.12	16) 176	

Converting Between Percent, Fractions, and Decimals

1) 0.5	7) 0.9	13) 52.4%
2) 0.38	8) 0.2	14) 10%
3) 0.15	9) 0.07	15) 3%
4) 0.8	10) 50%	16) 363%
5) 0.07	11) 90%	17) 0.8%
6) 0.35	12) 0.2%	18) 478%

Percent Problems

1) 15	7) 515.4	13) 49 hours
2) 104.3	8) 31.98	14) 62.7%
3) 38.34	9) 1.9%	15) Mathematics
4) 23.44%	10) 8.3 m	16) 600
5) 47.6%	11) 117 inches	
6) 100	12) 45.7 inches	

Find What Percentage a Number Is of Another

1) 45 is what percent of 90? 50 %

2) 15 is what percent of 75? 20 %

3) 20 is what percent of 400? 5 %

4) 18 is what percent of 90? 20 %

5) 3 is what percent of 15? 20 %

6) 8 is what percent of 80? 10 %

7) 11 is what percent of 55? 20 %

8) 9 is what percent of 90? 10 %

9) 2.5 is what percent of 10? 25 %

10) 5 is what percent of 25? 20 %

11) 60 is what percent of 20? 300 %

12) 12 is what percent of 48? 25 %

13) 14 is what percent of 28? 50 %

14) 8.2 is what percent of 32.8? 25 %

15) 1200 is what percent of 4,800?

    25 %

16) 4,000 is what percent of 20,000?

    20 %

17) 45 is what percent of 900? 5 %

18) 10 is what percent of 200? 5 %

19) 15 is what percent of 60? 25 %

20) 1.2 is what percent of 24? 5 %

Find a Percentage of a Given Number

1) 45	8) 3.5		15)	11.2	
2) 20	9) 8		16)	16	
3) 0	10)	6	17)	2.4	
4) 64	11)	50	18)	26.6	
5) 24	12)	30.6	19)	0.9	
6) 30	13)	33.6	20)	20	
7) 6	14)	10.8	21)	8	

Percent of Increase and Decrease

1) 156.25% increase

2) 70% decrease

3) 1.142% increase

4) 45.31% decrease

5) 15.75% increase

6) 150% increase

7) 92.6% increase

8) 71.43% decrease

9) 2.25% decrease

10) $28\frac{4}{7}$%

Test Preparation Answers

1) The answer is 60.

Jason needs an 75% average to pass for five exams. Therefore, the sum of 5 exams must be at least 5 × 75 = 375

The sum of 4 exams is: 68 + 72 + 85 + 90 = 315.

The minimum score Jason can earn on his fifth and final test to pass is: 375 − 315 = 60

2) Choice D is correct

To find the discount, multiply the number by (100% − rate of discount).

Therefore, for the first discount we get: (200) (100% − 15%) = (200) (0.85) = 170

For the next 15 % discount: (200) (0.85) (0.85)

3) Choice B is correct

Let x be the rate of depreciation.

If the price of a car is decreased by $4,000 to $16,000, then:

x % of $20.000 = $4.000 ⇒ $\frac{x}{100}$ 20.000 = 4.000 ⇒ x = 4.000 ÷ 200 = 20

4) Choice C is correct

Let x be the original price.

If the price of a laptop is decreased by 10% to $360, then:

$90\% \ of \ x = 360 \Rightarrow 0.90x = 360 \Rightarrow x = 360 \div 0.90 = 400$

5) Choice C is correct

Use simple interest formula: $I = prt$

(I = interest, p = principal, r = rate, t = time)

$$I = (8000)(0.045)(5) = 1800$$

6) The answer is 20%.

The whole number of students = 60 boys + 90 girls = 150

The number of students that write left-handed = 18 boys + 12 girls = 30

Let x be the percentage of the students that write left-handed.

$x\% \ of \ total \ students = \text{left} - \text{handed students} \Rightarrow x\%\ 150 = 30 \Rightarrow x = 3000 \div 150 = 20\%$

7) Choice C is correct

Use percent formula:

$\text{part} = \dfrac{\text{percent}}{100} \times \text{whole}$

$25 = \dfrac{\text{percent}}{100} \times 20 \Rightarrow 25 = \dfrac{\text{percent} \times 20}{100} \Rightarrow 25 = \dfrac{\text{percent} \times 2}{10}$, multiply both sides by 10.

250 = percent $\times$ 2, divide both sides by 2. 125 = percent

8) Choice B is correct

Use the formula for Percent of Change

$\dfrac{\text{New Value} - \text{Old Value}}{Old\ Value} \times 100\%$

$\dfrac{28-40}{40} \times 100\% = -30\%$ (negative sign here means that the new price is less than old price).

9) Choice C is correct

Use percent formula:

$$\text{part} = \frac{\text{percent}}{100} \times \text{whole}$$

$$1.75 = \frac{\text{percent}}{100} \times 1.25 \Rightarrow 1.75 = \frac{\text{percent} \times 1.25}{100} \Rightarrow 175 = \text{percent} \times 1.25 \Rightarrow$$

$$\text{percent} = \frac{175}{1.25} = 140$$

10) Choice B is correct.

The question is this: 530.40 is what percent of 624?

Use percent formula:

$$\text{part} = \frac{\text{percent}}{100} \times \text{whole}$$

$$530.40 = \frac{\text{percent}}{100} \times 624 \Rightarrow 530.40 = \frac{\text{percent} \times 624}{100} \Rightarrow 53040 = \text{percent} \times 624 \Rightarrow$$

$$\text{percent} = \frac{53040}{624} = 85$$

530.40 is 85 % of 624. Therefore, the discount is: 100% − 85% = 15%

11) Choice D is correct

The failing rate is 11 out of 55 $= \dfrac{11}{55}$

Change the fraction to percent:

$$\frac{11}{55} \times 100\% = 20\%$$

20 percent of students failed. Therefore, 80 percent of students passed the exam.

12) Choice C is correct.

Let x be the original price.

If the price of the sofa is decreased by 25% to $420, then: $75 \% \ of \ x = 420 \Rightarrow 0.75x = 420$ $\Rightarrow x = 420 \div 0.75 = 560$

Chapter 4: Algebraic Expressions

Topics that you'll learn in this chapter:

- ✓ Expressions and Variables
- ✓ Simplifying Variable Expressions
- ✓ Translate Phrases into an Algebraic Statement
- ✓ The Distributive Property
- ✓ Evaluating One Variable
- ✓ Evaluating Two Variables
- ✓ Combining like Terms

Expressions and Variables

Helpful	A variable is a letter that represents unknown numbers. A variable can be used in the same manner as all other numbers:		
Hints	Addition	$2 + a$	2 plus a
	Subtraction	$y - 3$	y minus 3
	Division	$\dfrac{4}{x}$	4 divided by x
	Multiplication	$5a$	5 times a

✍ *Simplify each expression.*

1) $x + 5x$,

   use $x = 5$

2) $8\,(-3x + 9) + 6$,

   use $x = 6$

3) $10x - 2x + 6 - 5$,

   use $x = 5$

4) $2x - 3x - 9$,

   use $x = 7$

5) $(-6)\,(-2x - 4y)$,

   use $x = 1$, $y = 3$

6) $8x + 2 + 4\,y$,

   use $x = 9$, $y = 2$

7) $(-6)\,(-8x - 9y)$,

   use $x = 5$, $y = 5$

8) $6x + 5y$,

   use $x = 7$, $y = 4$

✍ *Simplify each expression.*

9) $5\,(-4 + 2x)$

10) $-3 - 5x - 6x + 9$

11) $6x - 3x - 8 + 10$

12) $(-8)\,(6x - 4) + 12$

13) $9\,(7x + 4) + 6x$

14) $(-9)\,(-5x + 2)$

Simplifying Variable Expressions

Helpful Hints	– Combine "like" terms. (values with same variable and same power)	**Example:**
	– Use distributive property if necessary.	$2x + 2(1 - 5x) =$
	Distributive Property:	$2x + 2 - 10x = -8x + 2$
	$a(b + c) = ab + ac$	

✏️ *Simplify each expression.*

1) $-2 - x^2 - 6x^2$

2) $3 + 10x^2 + 2$
   $5 + 10x^2$

3) $8x^2 + 6x + 7x^2$

4) $5x^2 - 12x^2 + 8x$

5) $2x^2 - 2x - x$

6) $(-6)(8x - 4)$

7) $4x + 6(2 - 5x)$

8) $10x + 8(10x - 6)$

9) $9(-2x - 6) - 5$

10) $3(x + 9)$

11) $7x + 3 - 3x$

12) $2.5x^2 \times (-8x)$

✏️ *Simplify.*

13) $-2(4 - 6x) - 3x$, $x = 1$

14) $2x + 8x$, $x = 2$

15) $9 - 2x + 5x + 2$, $x = 5$

16) $5(3x + 7)$, $x = 3$

17) $2(3 - 2x) - 4$, $x = 6$

18) $5x + 3x - 8$, $x = 3$

19) $x - 7x$, $x = 8$

20) $5(-2 - 9x)$, $x = 4$

Translate Phrases into an Algebraic Statement

Helpful *Hints*	**Translating key words and phrases into algebraic expressions:**
	Addition: plus, more than, the sum of, etc.
	Subtraction: minus, less than, decreased, etc.
	Multiplication: times, product, multiplied, etc.
	Division: quotient, divided, ratio, etc.
	Example:
	eight more than a number is 20
	$8 + x = 20$

✎*Write an algebraic expression for each phrase.*

1) A number increased by forty–two.

2) The sum of fifteen and a number

3) The difference between fifty–six and a number.

4) The quotient of thirty and a number.

5) Twice a number decreased by 25.

6) Four times the sum of a number and – 12.

7) A number divided by – 20.

8) The quotient of 60 and the product of a number and – 5.

9) Ten subtracted from a number.

10) The difference of six and a number.

The Distributive Property

	Distributive Property:	Example:
Helpful	$a(b + c) = ab + ac$	$3(4 + 3x)$
Hints		$= 12 + 9x$

✍ **Use the distributive property to simply each expression.**

1) $-(-2 - 5x)$

2) $(-6x + 2)(-1)$

3) $(-5)(x - 2)$

4) $-(7 - 3x)$

5) $8(8 + 2x)$

6) $2(12 + 2x)$

7) $(-6x + 8)\,4$

8) $(3 - 6x)(-7)$

9) $(-12)(2x + 1)$

10) $(8 - 2x)\,9$

11) $(-2x)(-1 + 9x) - 4x(4 + 5x)$

12) $3(-5x - 3) + 4(6 - 3x)$

13) $(-2)(x + 4) - (2 + 3x)$

14) $(-4)(3x - 2) + 6(x + 1)$

15) $(-5)(4x - 1) + 4(x + 2)$

16) $(-3)(x + 4) - (2 + 3x)$

Evaluating One Variable

Helpful Hints	– To evaluate one variable expression, find the variable and substitute a number for that variable. – Perform the arithmetic operations.	**Example:** $4x + 8, x = 6$ $4(6) + 8 = 24 + 8 = 32$

✍️ *Simplify each algebraic expression.*

1) $9 - x$, $x = 3$

2) $x + 2$, $x = 5$

3) $3x + 7$, $x = 6$

4) $x + (-5)$, $x = -2$

5) $3x + 6$, $x = 4$

6) $4x + 6$, $x = -1$

7) $10 + 2x - 6$, $x = 3$

8) $10 - 3x$, $x = 8$

9) $\dfrac{20}{x} - 3$, $x = 5$

10) $(-3) + \dfrac{x}{4} + 2x$, $x = 16$

11) $(-2) + \dfrac{x}{7}$, $x = 21$

12) $(-\dfrac{14}{x}) - 9 + 4x$, $x = 2$

13) $(-\dfrac{6}{x}) - 9 + 2x$, $x = 3$

14) $(-2) + \dfrac{x}{8}$, $x = 16$

Evaluating Two Variables

To evaluate an algebraic expression, substitute a number for each variable and perform the arithmetic operations.

Example:

$$2x + 4y - 3 + 2,$$

$$x = 5, y = 3$$

$$2(5) + 4(3) - 3 + 2$$
$$= 10$$
$$+ 12 - 3 + 2$$
$$= 21$$

✎*Simplify each algebraic expression.*

1) $2x + 4y - 3 + 2,$

   $x = 5, y = 3$

2) $(-\dfrac{12}{x}) + 1 + 5y,$

   $x = 6, y = 8$

3) $(-4)(-2a - 2b),$

   $a = 5, b = 3$

4) $10 + 3x + 7 - 2y,$

   $x = 7, y = 6$

5) $9x + 2 - 4y,$

   $x = 7, y = 5$

6) $6 + 3(-2x - 3y),$

   $x = 9, y = 7$

7) $12x + y,$

   $x = 4, y = 8$

8) $x \times 4 \div y,$

   $x = 3, y = 2$

9) $2x + 14 + 4y,$

   $x = 6, y = 8$

10) $4a - (5 - b),$

   $a = 4, b = 6$

Combining like Terms

Helpful *Hints*	− Terms are separated by "+" and "−" signs. − Like terms are terms with same variables and same powers. − Be sure to use the "+" or "−" that is in front of the coefficient.	**Example:** $22x + 6 + 2x =$ $24x + 6$

✎ *Simplify each expression.*

1) $5 + 2x − 8$

2) $(− 2x + 6)\,2$

3) $7 + 3x + 6x − 4$

4) $(− 4) − (3)(5x + 8)$

5) $9x − 7x − 5$

6) $x − 12x$

7) $7\,(3x + 6) + 2x$

8) $(− 11x) − 10x$

9) $3x − 12 − 5x$

10) $13 + 4x − 5$

11) $(− 22x) + 8x$

12) $2\,(4 + 3x) − 7x$

13) $(− 4x) − (6 − 14x)$

14) $5\,(6x − 1) + 12x$

15) $22x + 6 + 2x$

16) $(− 13x) − 14x$

17) $(− 6x) − 9 + 15x$

18) $(− 6x) + 7x$

19) $(− 5x) + 12 + 7x$

20) $(− 3x) − 9 + 15x$

21) $20x − 19x$

Test Preparation

yucky

1) Which expression is equivalent to $38x$?

    A. $(x \times 30) \times 8$
    B. $(x \times 30) + 8$
    C. $(x \times 30) + (x \times 8)$
    D. $(x \times 3) + 8$

2) If $x = -8$, which equation is true?

    A. $x(2x - 4) = 120$ wrong
    B. $8(4 - x) = 96$ wrong
    C. $2(4x + 6) = 79$
    D. $6x + 2 = -50$

Answers of Worksheets – Chapter 4

Expressions and Variables

1) 30	6) 82	11) $3x + 2$
2) −66	7) 510	12) $44 − 48x$
3) 41	8) 62	13) $69x + 36$
4) −16	9) $10x − 20$	14) $45x − 18$
5) 84	10) $6 − 11x$	

Simplifying Variable Expressions

1) $-7x^2 - 2$	8) $90x − 48$	15) 26
2) $10x^2 + 5$	9) $-18x − 59$	16) 80
3) $15x^2 + 6x$	10) $3x + 27$	17) -22
4) $-7x^2 + 8x$	11) $4x + 3$	18) 16
5) $2x^2 − 3x$	12) $-20x^3$	19) -48
6) $-48x + 24$	13) 1	20) -190
7) $-26x + 12$	14) 20	

Translate Phrases into an Algebraic Statement

1) $x + 42$	2) $15 + x$
3) $56 − x$	6) $4(x + (−12))$
4) $30/x$	7) $\dfrac{x}{-20}$
5) $2x − 25$	
8) $\dfrac{60}{-5x}$	10) $6 − x$
9) $x − 10$	

The Distributive Property

1) $5x + 2$	7) $-24x + 32$	13) $-5x − 10$
2) $6x − 2$	8) $42x − 21$	14) $-6x + 14$
3) $-5x + 10$	9) $-24x − 12$	15) $-16x + 13$
4) $3x − 7$	10) $-18x + 72$	16) $-6x − 14$
5) $16x + 64$	11) $-38x^2 − 14x$	
6) $4x + 24$	12) $-27x + 15$	

Evaluating One Variable

1) 6
2) 7
3) 25
4) −7
5) 18

6) 2
7) 10
8) −14
9) 1
10) 33

11) 1
12) −8
13) −5
14) 0
15) −176

Evaluating Two Variables

1) 21
2) 39
3) 64
4) 26

5) 45
6) −111
7) 56
8) 6

9) 58
10) 17

Combining like Terms

1) $2x - 3$
2) $-4x + 12$
3) $9x + 3$
4) $-15x - 28$
5) $2x - 5$
6) $-11x$
7) $23x + 42$

8) $-21x$
9) $-2x - 12$
10) $4x + 8$
11) $-14x$
12) $-x + 8$
13) $10x - 6$
14) $42x - 5$

15) $24x + 6$
16) $-27x$
17) $9x - 9$
18) x
19) $2x + 12$
20) $12x - 9$
21) x

Test Preparation Answers

1) Choice C is correct.

Let's can alternatives on by one.

A. $(30x) \times 8 = 240x$

B. $(30x) + 8 = 30x + 8$

C. $(30x) + (8x) = 38x$

D. $(3x) + 8 = 3x + 8$

2) Choice B is correct.

$x = -8$, then:

A. $(-8)(2(-8) - 4) = 120 \rightarrow 160 = 120$ Wrong!

B. $8(4 - (-8)) = 96 \rightarrow 96 = 96$ Correct!

C. $2(4(-8) + 6) = 79 \rightarrow -52 = 79$ Wrong!

D. $6(-8) + 2 = -50 \rightarrow -46 = -50$ Wrong!

Chapter 5: Equations

Topics that you'll learn in this chapter:

- ✓ One–Step Equations
- ✓ One–Step Equation Word Problems
- ✓ Two–Step Equations
- ✓ Two–Step Equation Word Problems
- ✓ Multi–Step Equations

One–Step Equations

Helpful	-	The values of two expressions on both sides of an equation are equal.	**Example:**

$$ax + b = c$$

Hints	-	You only need to perform one Math operation in order to solve the equation.	$-8x = 16$

$$x = -2$$

✎**Solve each equation.**

1) x + 3 = 17

    x = 14

2) 22 = (− 8) + x

    x = 30

3) 3x = (− 30)

    x = -10

4) (− 36) = (− 6x)

    x = 6

5) (− 6) = 4 + x

    x = -10

6) 2 + x = (− 2)

    x = -4

7) 20x = (− 220)

    x = -11

8) 18 = x + 5

    x = 13

9) (− 23) + x = (− 19)

    x = 4

10) 5x = (− 45)

    x = -9

11) x − 12 = (− 25)

    x = 37

12) x − 3 = (− 12)

    x = -15

13) (− 35) = x − 27

    x = 62

14) 8 = 2x

    x = 4

15) (− 6x) = 36

    x = 6

16) (− 55) = (− 5x)

    x = 11

17) x − 30 = 20

    x = 50

18) 8x = 32

    x = 4

19) 36 = (− 4x)

    x = 9

20) 4x = 68

    x = 17

21) 30x = 300

    x = 10

One–Step Equation Word Problems

Helpful *Hints*	– Define the variable.
	– Translate key words and phrases into math equation.
	– Isolate the variable and solve the equation.

✎ **Solve.**

1) How many boxes of envelopes can you buy with $18 if one box costs $3?

    $3x = 18$ $x = 6$

2) After paying $6.25 for a salad, Ella has $45.56. How much money did she have before buying the salad?

    $x - 6.25 = 45.56$ $x = 39.31$

    $$\begin{array}{r} 45.56 \\ 6.25 \\ \hline 39.31 \end{array}$$

3) How many packages of diapers can you buy with $50 if one package costs $5?

    $5x = 50$
    $x = 10$

4) Last week James ran 20 miles more than Michael. James ran 56 miles. How many miles did Michael run?

    $x + 20 = 56$
    $x = 36$

5) Last Friday Jacob had $32.52. Over the weekend he received some money for cleaning the attic. He now has $44. How much money did he receive?

    $32.52 + x = 44$ $x = 11.48$

    $$\begin{array}{r} 44.00 \\ 32.52 \\ \hline 11.48 \end{array}$$

6) After paying $10.12 for a sandwich, Amelia has $35.50. How much money did she have before buying the sandwich?

    $x - 10.12 = 35.50$ $x = 25.38$

    $$\begin{array}{r} 35.50 \\ 10.12 \\ \hline 25.38 \end{array}$$

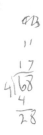

Two–Step Equations

Helpful	– You only need to perform two math operations (add, subtract, multiply, or divide) to solve the equation.	**Example:**
Hints	– Simplify using the inverse of addition or subtraction.	$-2(x-1) = 42$ $(x-1) = -21$ $x = -20$
	– Simplify further by using the inverse of multiplication or division.	

✍ *Solve each equation.*

1) $5(8 + x) = 20$

2) $(-7)(x - 9) = 42$

3) $(-12)(2x - 3) = (-12)$

4) $6(1 + x) = 12$

5) $12(2x + 4) = 60$

6) $7(3x + 2) = 42$

7) $8(14 + 2x) = (-34)$

8) $(-15)(2x - 4) = 48$

9) $3(x + 5) = 12$

10) $\dfrac{3x - 12}{6} = 4$

11) $(-12) = \dfrac{x + 15}{6}$

12) $110 = (-5)(2x - 6)$

13) $\dfrac{x}{8} - 12 = 4$

14) $20 = 12 + \dfrac{x}{4}$

15) $\dfrac{-24 + x}{6} = (-12)$

16) $(-4)(5 + 2x) = (-100)$

17) $(-12x) + 20 = 32$

18) $\dfrac{-2 + 6x}{4} = (-8)$

19) $\dfrac{x + 6}{5} = (-5)$

20) $(-9) + \dfrac{x}{4} = (-15)$

Two–Step Equation Word Problems

Helpful *Hints*	– Translate the word problem into equations with variables. – Solve the equations to find the solutions to the word problems.

✎ **Solve.**

1) The sum of three consecutive even numbers is 48. What is the smallest of these numbers?

2) How old am I if 400 reduced by 2 times my age is 244?

$$18$$

$$400 - 2x = 244$$

$$2x =$$

$$\begin{array}{r} 78 \\ 2\overline{)156} \\ 14 \\ \hline 16 \end{array}$$

$$\begin{array}{r} 400 \\ 244 \\ \hline 156 \end{array}$$

3) For a field trip, 4 students rode in cars and the rest filled nine buses. How many students were in each bus if 472 students were on the trip?

$$x = 52$$

$$4 + 9x = 472$$

$$9x = 468$$

$$\begin{array}{r} 52 \\ 9\overline{)468} \\ 45 \\ \hline 18 \end{array}$$

4) The sum of three consecutive numbers is 72. What is the smallest of these numbers?

5) 331 students went on a field trip. Six buses were filled, and 7 students traveled in cars. How many students were in each bus?

$$6x + 7 = 331$$

$$6x = 324$$

$$x = 54$$

$$\begin{array}{r} 54 \\ 6\overline{)324} \\ 300 \\ \hline 24 \end{array}$$

6) You bought a magazine for $5 and four erasers. You spent a total of $25. How much did each eraser cost?

$$5 + 4x = 25 \qquad x = 5$$

Multi–Step Equations

Helpful *Hints*	– Combine "like" terms on one side. – Bring variables to one side by adding or subtracting. – Simplify using the inverse of addition or subtraction. – Simplify further by using the inverse of multiplication or division.	**Example:** $3x + 15 = -2x + 5$ Add 2x both sides $5x + 15 = +5$ Subtract 15 both sides $5x = -10$ Divide by 5 both sides $x = -2$

Solve each equation.

1) $-(2 - 2x) = 10$

2) $-12 = -(2x + 8)$

3) $3x + 15 = (-2x) + 5$

4) $-28 = (-2x) - 12x$

5) $2(1 + 2x) + 2x = -118$

6) $3x - 18 = 22 + x - 3 + x$

7) $12 - 2x = (-32) - x + x$

8) $7 - 3x - 3x = 3 - 3x$

9) $6 + 10x + 3x = (-30) + 4x$

10) $(-3x) - 8(-1 + 5x) = 352$

11) $24 = (-4x) - 8 + 8$

12) $9 = 2x - 7 + 6x$

13) $6(1 + 6x) = 294$

14) $-10 = (-4x) - 6x$

15) $4x - 2 = (-7) + 5x$

16) $5x - 14 = 8x + 4$

17) $40 = -(4x - 8)$

18) $(-18) - 6x = 6(1 + 3x)$

19) $x - 5 = -2(6 + 3x)$

20) $6 = 1 - 2x + 5$

Test Preparation

1) What is the value of x in the following equation?

$$x + \frac{1}{6} = \frac{1}{3}$$

$$X + \frac{2}{12} = \frac{4}{12}$$

   A. 6

   B. $\frac{1}{2}$

   C. $\frac{1}{6}$

   D. $\frac{1}{4}$

2) The area of a rectangle is x square feet and its length is 9 feet. Which equation represents y, the width of the rectangle in feet?

   A. $y = \frac{x}{9}$

   B. $y = \frac{9}{x}$

   C. $y = 9x$

   D. $y = 9 + x$

3) An angle is equal to one fifth of its supplement. What is the measure of that angle?

A. 20
B. 30
C. 45
D. 60

4) In five successive hours, a car travels 40 km, 45 km, 50 km, 35 km and 55 km. In the next five hours, it travels with an average speed of 50 km per hour. Find the total distance the car traveled in 10 hours.

A. 425 km
B. 450 km
C. 475 km
D. 500 km

5) In a triangle ABC the measure of angle ACB is 75° and the measure of angle CAB is 45°. What is the measure of angle ABC?

Write your answer in the box below.

6) If 40 % of a number is 4, what is the number?

    A. 4
    B. 8
    C. 10
    D. 12

7) Jason is 9 miles ahead of Joe running at 5.5 miles per hour and Joe is running at the speed of 7 miles per hour. How long does it take Joe to catch Jason?

    A. 3 hours
    B. 4 hours
    C. 6 hours
    D. 8 hours

8) What is the equivalent temperature of 104°F in Celsius?

$$C = \frac{5}{9}(F - 32)$$

    A. 32
    B. 40
    C. 48
    D. 52

9) If 150 % of a number is 75, then what is the 90 % of that number?

    A. 45

    B. 50

    C. 70

    D. 85

Answers of Worksheets – Chapter 5

One–Step Equations

1) 14
2) 30
3) – 10
4) 6
5) – 10
6) – 4
7) – 11

8) 13
9) 4
10) – 9
11) – 13
12) – 9
13) – 8
14) 4

15) – 6
16) 11
17) 50
18) 4
19) – 9
20) 17
21) 10

One–Step Equation Word Problems

1) 6
2) $51.81

3) 10
4) 36

5) 11.48
6) 45.62

Two–Step Equations

1) – 4
2) 3
3) 2
4) 1
5) 0.5
6) $\frac{4}{3}$
7) $-\frac{73}{8}$

8) $\frac{2}{5}$
9) – 1
10) 12
11) – 87
12) – 8
13) 128
14) 32

15) – 48
16) 10
17) – 1
18) – 5
19) – 31
20) – 24

Two–Step Equation Word Problems

1) 14
2) 78

3) 52
4) 23

5) 54
6) $4

Multi–Step Equations

1) 6
2) 2
3) − 2
4) 2
5) − 20
6) 37
7) 22

8) $\dfrac{4}{3}$
9) − 4
10) − 8
11) − 6
12) 2
13) 8

14) 1
15) 5
16) − 6
17) − 8
18) − 1
19) − 1
20) 0

Test Preparation Answers

1) The answer is C.

Subtract the numerators and find x.

$$x = \frac{1}{3} - \frac{1}{6} \Rightarrow x = \frac{2-1}{6} \Rightarrow x = \frac{1}{6}$$

2) Choice A is correct

To find the area of rectangle = width multiply length

Therefore; $x = 9 \times y$

Then find y : $y = \frac{x}{9}$

3) Choice B is correct

The sum of supplement angles is 180. Let x be that angle. Therefore,

$x + 5x = 180$

$6x = 180$, divide both sides by 6: $x = 30$

4) Choice C is correct

Add the first 5 numbers. 40 + 45 + 50 + 35 + 55 = 225

To find the distance traveled in the next 5 hours, multiply the average by number of hours.

Distance = Average × Rate = 50 × 5 = 250

Add both numbers.

250 + 225 = 475

5) The answer is 60.

The whole angles in every triangle are:180°and Let x be the number of new angle so:

$180 = 75 + 45 + x \Rightarrow x = 60°$

6) Choice C is correct

Let x be the number. Write the equation and solve for x.

40% of $x = 4 \Rightarrow 0.40\, x = 4 \Rightarrow x = 4 \div 0.40 = 10$

7) Choice C is correct

The distance between Jason and Joe is 9 miles. Jason running at 5.5 miles per hour and Joe is running at the speed of 7 miles per hour. Therefore, every hour the distance is 1.5 miles less. 9 ÷ 1.5 = 6

8) Choice B is correct

Plug in 104 for F and then solve for C.

$$C = \frac{5}{9}\,(F - 32) \Rightarrow C = \frac{5}{9}\,(104 - 32) \Rightarrow C = \frac{5}{9}\,(72) = 40$$

9) Choice A is correct

First, find the number.

Let x be the number. Write the equation and solve for x.

150 % of a number is 75, then:

$1.5 \times x = 75 \Rightarrow x = 75 \div 1.5 = 50$

90 % of 50 is:

0.9 × 50 = 45

Chapter 6: Proportions and Ratios

Topics that you'll learn in this chapter:

- ✓ Writing Ratios
- ✓ Simplifying Ratios
- ✓ Proportional Ratios
- ✓ Create a Proportion
- ✓ Similar Figures
- ✓ Similar Figure Word Problems
- ✓ Ratio and Rates Word Problems

Writing Ratios

Helpful	− A ratio is a comparison of two numbers. Ratio can be written as a division.	**Example:**
Hints		$3:5$, or $\frac{3}{5}$

✏️ **Express each ratio as a rate and unite rate.**

1) 120 miles on 4 gallons of gas.

   120:4

2) 24 dollars for 6 books.

   24:6

3) 200 miles on 14 gallons of gas

   200:14

4) 24 inches of snow in 8 hours

   24:8

✏️ **Express each ratio as a fraction in the simplest form.**

5) 3 feet out of 30 feet
   $\frac{1}{10}$

6) 18 cakes out of 42 cakes
   $\frac{3}{7}$

7) 16 dimes t0 24 dimes
   $\frac{2}{3}$

8) 12 dimes out of 48 coins
   $\frac{1}{4}$

9) 14 cups to 84 cups
   $\frac{7}{12}$

10) 45 gallons to 65 gallons
    $\frac{9}{13}$

11) 10 miles out of 40 miles
    $\frac{1}{4}$

12) 22 blue cars out of 55 cars
    $\frac{2}{5}$

13) 32 pennies to 300 pennies
    $\frac{8}{75}$

14) 24 beetles out of 86 insects
    $\frac{12}{43}$

Proportional Ratios

> **Helpful Hints**
>
> – A proportion means that two ratios are equal. It can be written in two ways:
>
> $$\frac{a}{b} = \frac{c}{d}, \quad a : b = c : d$$
>
> **Example:**
>
> $$\frac{9}{3} = \frac{6}{d}$$
>
> $d = 2$

2.66
6

Solve each proportion.

2.6
6
15.6

1) $\frac{3}{6} = \frac{8}{d}$ 15

2 86
3⟌8
6
20

8) $\frac{60}{20} = \frac{3}{d}$ 1

15) $\frac{x}{4} = \frac{30}{5}$ 24

2) $\frac{k}{5} = \frac{12}{15}$ 4

9) $\frac{x}{3} = \frac{12}{18}$ 2

16) $\frac{9}{5} = \frac{k}{5}$ 9

3) $\frac{30}{5} = \frac{12}{x}$ 2

$\frac{30}{12} = \frac{5}{2} = 2\frac{1}{2}$

2.5
5
12.5

10) $\frac{25}{5} = \frac{x}{8}$ 40

$3 : \frac{9}{3}$

17) $\frac{45}{15} = \frac{15}{d}$ 5

4) $\frac{x}{2} = \frac{1}{8}$ .25

2
8⟌2.8

11) $\frac{12}{x} = \frac{4}{2}$ 6

18) $\frac{60}{x} = \frac{10}{3}$ .5

5) $\frac{d}{3} = \frac{2}{6}$ 1

12) $\frac{x}{4} = \frac{18}{2}$ 36

19) $\frac{d}{3} = \frac{14}{6}$ 7

6) $\frac{27}{7} = \frac{30}{x}$ 7.78

3.5
7⟌27
2
80
4

13) $\frac{80}{10} = \frac{k}{10}$ 80

20) $\frac{k}{4} = \frac{4}{2}$ 2

7) $\frac{8}{5} = \frac{k}{15}$ 24

14) $\frac{12}{6} = \frac{6}{d}$ 3

3.5
7
2
15.0

21) $\frac{4}{2} = \frac{x}{7}$ 14

Simplifying Ratios

Helpful *Hints*	– You can calculate equivalent ratios by multiplying or dividing both sides of the ratio by the same number.	**Examples:** 3 : 6 = 1 : 2 4 : 9 = 8 : 18

✎ *Reduce each ratio.*

1) 21 : 49
   3:7

2) 20 : 40
   1:2

3) 10: 50
   1:5

4) 14: 18
   7:9

5) 45: 27
   5:3

6) 49: 21
   7:3

7) 100: 10
   10:1

8) 12 : 8
   3:2

9) 35 : 45
   7:9

10) 8: 20
    2:5

11) 25: 35
    5:7

12) 21 : 27
    7:9

13) 52 : 82
    26:41

14) 12: 36
    1:3

15) 24 : 3
    8:1

16) 15: 30
    1:2

17) 3 : 36
    1:12

18) 8 : 16
    1:2

19) 6 : 100
    3:50

20) 2 : 20
    1:10

21) 10: 60
    1:6

22) 14: 63
    2:9

23) 68: 80
    17:20

24) 8: 80
    1:10

Create a Proportion

Helpful Hints	– A proportion contains 2 equal fractions! A proportion simply means that two fractions are equal.	**Example:** 2, 4, 8, 16 $$\frac{2}{4} = \frac{8}{16}$$

✍ Create proportion from the given set of numbers.

1) 1, 6, 2, 3
$$\frac{1}{2} = \frac{3}{6}$$

2) 12, 144, 1, 12
$$\frac{1}{12} = \frac{12}{144}$$

3) 16, 4, 8, 2
$$\frac{2}{4} = \frac{8}{16}$$

4) 9, 5, 27, 15
$$\frac{5}{9} = \frac{15}{27}$$

5) 7, 10, 60, 42
$$\frac{7}{10} = \frac{42}{60}$$

6) 8, 7, 24, 21
$$\frac{7}{8} = \frac{21}{24}$$

7) 10, 5, 8, 4
$$\frac{4}{5} = \frac{8}{10}$$

8) 3, 12, 8, 2
$$\frac{2}{3} = \frac{8}{12}$$

9) 2, 2, 1, 4
$$\frac{1}{2} = \frac{2}{4}$$

10) 3, 6, 7, 14
$$\frac{3}{6} = \frac{7}{14}$$

11) 2, 6, 5, 15
$$\frac{2}{5} = \frac{6}{15}$$

12) 7, 2, 14, 4
$$\frac{2}{4} = \frac{7}{14}$$

Similar Figures

Helpful Hints	– Two or more figures are similar if the corresponding angles are equal, and the corresponding sides are in proportion.	**Example:** 3–4–5 triangle is similar to a 6–8–10 triangle

✏ *Each pair of figures is similar. Find the missing side.*

1)

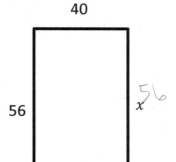

12

15

4

x 5

$4\overline{)60}$

2)

$X=3$

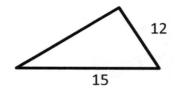

5x

8

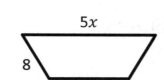

60

32

3)

40

56

x 56

7

5

5

7

Similar Figure Word Problems

Helpful

Hints

To solve a similarity word problem, create a proportion and use cross multiplication method!

Example:

$$\frac{x}{4} = \frac{8}{16}$$

$$16x = 4 \times 8$$

$$x = 2$$

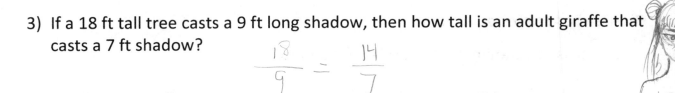

✎ *Answer each question and round your answer to the nearest whole number.*

1) If a 42.9 ft tall flagpole casts a 253.1 ft long shadow, then how long is the shadow that a 6.2 ft tall woman casts?

$$\frac{42.9}{253.1} = \frac{6.2}{36.6}$$

2) A model igloo has a scale of 1 in : 2 ft. If the real igloo is 10 ft wide, then how wide is the model igloo?

$$\frac{1}{24} = \frac{5}{120}$$

3) If a 18 ft tall tree casts a 9 ft long shadow, then how tall is an adult giraffe that casts a 7 ft shadow?

$$\frac{18}{9} = \frac{14}{7}$$

4) Find the distance between San Joe and Mount Pleasant if they are 2 cm apart on a map with a scale of 1 cm : 9 km.

18

5) A telephone booth that is 8 ft tall casts a shadow that is 4 ft long. Find the height of a lawn ornament that casts a 2 ft shadow.

4

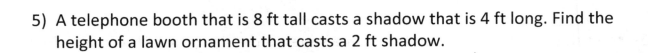

Ratio and Rates Word Problems

Helpful *Hints*	To solve a ratio or a rate word problem, create a proportion and use cross multiplication method!	**Example:** $\dfrac{x}{4} = \dfrac{8}{16}$ $16x = 4 \times 8$ $x = 2$

✎Solve.

$\dfrac{21}{12\overline{)252}}$
$\dfrac{24}{12}$

1) In a party, 10 soft drinks are required for every 12 guests. If there are 252 guests, how many soft drink is required?   210

2) In Jack's class, 18 of the students are tall and 10 are short. In Michael's class 54 students are tall and 30 students are short. Which class has a higher ratio of tall to short students?   9:5   they're equal   9:5

$\dfrac{12}{72}$
3) Are these ratios equivalent?

   12 cards to 72 animals 11 marbles to 66 marbles
   1:6   Yes   1:6

$\dfrac{48}{3\overline{)144}}$
$\dfrac{12}{24}$
4) The price of 3 apples at the Quick Market is \$1.44. The price of 5 of the same apples at Walmart is \$2.50. Which place is the better buy?
   0.48   Quick Mart   .5 $5\overline{)2.5}$

5) The bakers at a Bakery can make 160 bagels in 4 hours. How many bagels can they bake in 16 hours? What is that rate per hour? $\dfrac{160}{4}$ 640 $\dfrac{2}{640}$

6) You can buy 5 cans of green beans at a supermarket for \$3.40. How much does it cost to buy 35 cans of green beans?   3.4   22.8

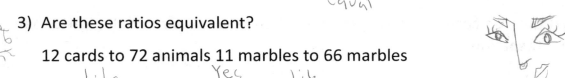

Test Preparation

1) The ratio of boys and girls in a class is 4:7. If there are 44 students in the class, how many more boys should be enrolled to make the ratio 1:1?

   A. 8

   B. 10

   C. 12

   D. 14

2) In a party, 6 soft drinks are required for every 9 guests. If there are 171 guests, how many soft drink is required?

   A. 9

   B. 27

   C. 114

   D. 171

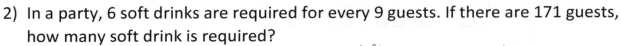

3) Peter traveled 150 km in 6 hours and Jason traveled 180 km in 4 hours. What is the ratio of the average speed of Peter to average speed of Jason?

   A. 3 : 2

   B. 2 : 3

   C. 5 : 9

   D. 5 : 6

4) How long does a 420–miles trip take moving at 50 miles per hour (mph)?

A. 4 hours
B. 6 hours and 24 minutes
C. 8 hours and 24 minutes
D. 8 hours and 30 minutes

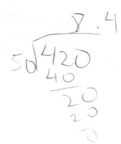

5) The ratio of boys to girls in a school is 3:4. If there are 350 students in a school, how many boys are in the school.

Write your answer in the box below.

200

Answers of Worksheets – Chapter 6

Writing Ratios

1) $\frac{120 \text{ miles}}{4 \text{ gallons}}$, 30 miles per gallon

2) $\frac{24 \text{ dollars}}{6 \text{ books}}$, 4.00 dollars per book

3) $\frac{200 \text{ miles}}{14 \text{ gallons}}$, 14.29 miles per gallon

4) $\frac{24" \text{ of snow}}{8 \text{ hours}}$, 3 inches of snow per hour

5) $\frac{1}{10}$

6) $\frac{3}{7}$

7) $\frac{2}{3}$

8) $\frac{1}{4}$

9) $\frac{1}{6}$

10) $\frac{9}{13}$

11) $\frac{1}{4}$

12) $\frac{2}{5}$

13) $\frac{8}{75}$

14) $\frac{12}{43}$

Simplifying Ratios

1) 3 : 7
2) 1 : 2
3) 1 : 5
4) 7 : 9
5) 5 : 3
6) 7 : 3
7) 10 : 1
8) 3 : 2

9) 7 : 9
10) 2 : 5
11) 5 : 7
12) 7 : 9
13) 26 : 41
14) 1 : 3
15) 8 : 1
16) 1 : 2

17) 1 : 12
18) 1 : 2
19) 3 : 50
20) 1 : 10
21) 1 : 6
22) 2 : 9
23) 17 : 20
24) 1 : 10

Proportional Ratios

1) 16
2) 4
3) 2
4) 0.25

5) 1
6) 7.78
7) 24
8) 1

9) 2
10) 40
11) 6
12) 36

13) 80

14) 3

15) 24

16) 9

17) 5

18) 18

19) 7

20) 8

21) 14

Create a Proportion

1) $1 : 3 = 2 : 6$

2) $12 : 144 = 1 : 12$

3) $2 : 4 = 8 : 16$

4) $5 : 15 = 9 : 27$

5) $7 : 42, 10 : 60$

6) $7 : 21 = 8 : 24$

7) $8 : 10 = 4 : 5$

8) $2 : 3 = 8 : 12$

9) $4 : 2 = 2 : 1$

10) $7 : 3 = 14 : 6$

11) $5 : 2 = 15 : 6$

12) $7 : 2 = 14 : 4$

Similar Figures

1) 5

2) 3

3) 56

Similar Figure Word Problems

1) 36.6 ft

2) 5 in

3) 14 ft

4) 18 km

5) 4 ft

Ratio and Rates Word Problems

1) 210

2) The ratio for both class is equal to 9 to 5.

3) Yes! Both ratios are 1 to 6

4) The price at the Quick Market is a better buy.

5) 640, the rate is 40 per hour.

6) $23.80

Test Preparation Answers

1) Choice C is correct

The ratio of boy to girls is 4:7. Therefore, there are 4 boys out of 11 students. To find the answer, first divide the total number of students by 11, then multiply the result by 4.

$44 \div 11 = 4 \Rightarrow 4 \times 4 = 16$

There are 16 boys and 28 (44 − 16) girls. So, 12 more boys should be enrolled to make the ratio 1:1

2) Choice C is correct

Let x be the number of soft drinks for 171 guests. It's needed to have a proportional ratio to find x.

$$\frac{6 \text{ soft drinks}}{9 \text{ guests}} = \frac{x}{171 \text{ guests}}$$

$$x = \frac{171 \times 6}{9} \Rightarrow x = 114$$

3) Choice C is correct

$$\text{Speed} = \frac{The\ amount\ of\ Km}{The\ amount\ of\ hours}$$

Peter's speed $= \dfrac{150}{6} = 25$

Jason's speed $= \dfrac{180}{4} = 45$

$\dfrac{The\ average\ speed\ of\ peter}{The\ average\ speed\ of\ Jason} = \dfrac{25}{45}$ and after simplification we have: $\dfrac{5}{9}$

4) Choice C is correct

Use distance formula:

Distance = Rate × time $\Rightarrow$ 420 = 50 × T, divide both sides by 50. 420 / 50 = T $\Rightarrow$ T = 8.4 hours.

Change hours to minutes for the decimal part. 0.4 hours = 0.4 × 60 = 24 minutes.

5) The answer is 150.

The ratio of boy to girls is 3:4. Therefore, there are 3 boys out of 7 students. To find the answer, first divide the total number of students by 7, then multiply the result by 3.

$350 \div 7 = 50 \Rightarrow 50 \times 3 = 150$

Chapter 7: Inequalities

Topics that you'll learn in this chapter:

- ✓ Graphing Single– Variable Inequalities
- ✓ One– Step Inequalities
- ✓ Two– Step Inequalities
- ✓ Multi– Step Inequalities

Graphing Single–Variable Inequalities

Helpful *Hints*	– Isolate the variable. – Find the value of the inequality on the number line. – For less than or greater than draw open circle on the value of the variable. – If there is an equal sign too, then use filled circle. – Draw a line to the right direction.

✍ *Draw a graph for each inequality.*

1) $-2 > x$

2) $5 \leq -x$

3) $x > 7$

4) $-x > 1.5$

One–Step Inequalities

Helpful *Hints*	– Isolate the variable. – For dividing both sides by negative numbers, flip the direction of the inequality sign.	**Example:** $x + 4 \geq 11$ $x \geq 7$

✎ *Solve each inequality and graph it.*

1) $x + 9 \geq 11$

$X \geq 2$

2) $x - 4 \leq 2$

$X \leq 6$

3) $6x \geq 36$

$X \geq 1$

4) $7 + x < 16$

$X < 9$

5) $x + 8 \leq 1$

$X \leq -7$

6) $3x > 12$

$X > 4$

7) $3x < 24$

$X < 8$

Two–Step Inequalities

Helpful	– Isolate the variable.	**Example:**
Hints	– For dividing both sides by negative numbers, flip the direction of the of the inequality sign.	$2x + 9 \geq 11$ $2x \geq 2$
	– Simplify using the inverse of addition or subtraction.	$x \geq 1$
	– Simplify further by using the inverse of multiplication or division.	

✍ *Solve each inequality and graph it.*

1) $3x - 4 \leq 5$

$3x \leq 9$

$x \leq 3$

2) $2x - 2 \leq 6$

$2x \leq 8$

$x \leq 4$

3) $4x - 4 \leq 8$

$x \leq 3$

4) $3x + 6 \geq 12$

$x \geq 2$

5) $6x - 5 \geq 19$

$x \geq 4$

6) $2x - 4 \leq 6$

$x \leq 5$

7) $8x - 4 \leq 4$

$x \leq 1$

8) $6x + 4 \leq 10$

$x \leq 1$

9) $5x + 4 \leq 9$

$x \leq 1$

10) $7x - 4 \leq 3$

$x \leq 1$

11) $4x - 19 < 19$

$x < 9.5$

12) $2x - 3 < 21$

$x < 12$

13) $7 + 4x \geq 19$

$x \geq 3$

14) $9 + 4x < 21$

$x < 3$

15) $3 + 2x \geq 19$

$x \geq 8$

16) $6 + 4x < 22$

$x < 4$

Multi–Step Inequalities

Helpful *Hints*	– Isolate the variable. – Simplify using the inverse of addition or subtraction. – Simplify further by using the inverse of multiplication or division.	**Example:** $\dfrac{7x+1}{3} \geq 5$ $7x+1 \geq 15$ $7x \geq 14$ $x \geq 7$

✍ *Solve each inequality.*

1) $\dfrac{9x}{7} - 7 < 2$

2) $\dfrac{4x+8}{2} \leq 12$

3) $\dfrac{3x-8}{7} > 1$

4) $-3(x-7) > 21$

5) $4 + \dfrac{x}{3} < 7$

6) $\dfrac{2x+6}{4} \leq 10$

Test Preparation

1) A football team had $20,000 to spend on supplies. The team spent $14,000 on new balls. New sport shoes cost $120 each. Which of the following inequalities represent how many new shoes the team can purchase.

    A. $120x + 14,000 \leq 20,000$

    B. $120x + 14,000 \geq 20,000$

    C. $14,000x + 120 \leq 20,000$

    D. $14,000x + 120 \geq 20,000$

2) In 1999, the average worker's income increased $2,000 per year starting from $24,000 annual salary. Which equation represents income greater than average? (I = income, x = number of years after 1999)

    A. $I > 2000\ x + 24000$

    B. $I > -\ 2000\ x + 24000$

    C. $I < -2000\ x + 24000$

    D. $I < 2000\ x - 24000$

3) Which of the following graphs represents the compound inequality $-2 \le 2x - 4 < 8$?

A.

B.

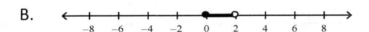

C.

D.

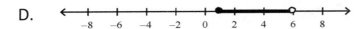

Answers of Worksheets – Chapter 7

Graphing Single–Variable Inequalities

1) $-2 > x$

2) $x \leq -5$

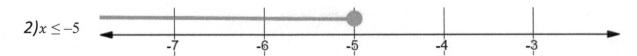

3) $x > 7$

4) $-1.5 > x$

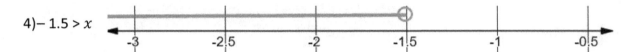

One–Step Inequalities

1)

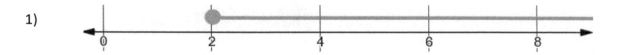

2)

3)

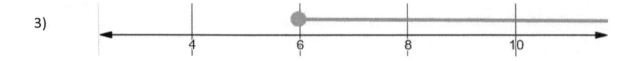

4)

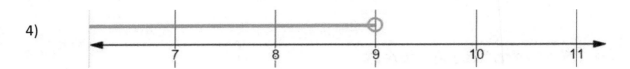

5)

6)

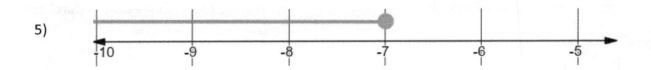

7)

Two–Step inequalities

1) $x \le 3$

2) $x \le 4$

3) $x \le 3$

4) $x \ge 2$

5) $x \ge 4$

6) $x \le 5$

7) $x \le 1$

8) $x \le 1$

9) $x \le 1$

10) $x \le 1$

11) $x < 9.5$

12) $x < 12$

13) $x \ge 3$

14) $x < 3$

15) $x \ge 8$

16) $x < 4$

Multi–Step inequalities

1) $x < 7$

2) $x \le 4$

3) $x > 5$

4) $x < 0$

5) $x < 9$

6) $x \le 17$

Test Preparation Answers

1) Choice A is correct

Let x be the number of new shoes the team can purchase. Therefore, the team can purchase $120\,x$.

The team had $20,000 and spent $14000. Now the team can spend on new shoes $6000 at most.

Now, write the inequality:

$$120x + 14.000 \leq 20.000$$

2) Choice A is correct

Let x be the number of years. Therefore, $2,000 per year equals $2000x$.

starting from $24,000 annual salary means you should add that amount to $2000x$.

Income more than that is:

$I > 2000x + 24000$

3) Choice D is correct

$-2 \leq 2x - 4 < 8 \rightarrow$ (add 4 all sides) $-2 + 4 \leq 2x < 8 + 4 \rightarrow 2 \leq 2x < 12$
$\rightarrow$ (divide all sides by 2) $1 \leq x < 6$

Chapter 8: Exponents and Radicals

Topics that you'll learn in this chapter:

✓ Multiplication Property of Exponents

✓ Division Property of Exponents

✓ Powers of Products and Quotients

✓ Zero and Negative Exponents

✓ Negative Exponents and Negative Bases

✓ Writing Scientific Notation

✓ Square Roots

Multiplication Property of Exponents

Helpful *Hints*	**Exponents rules** $x^a \cdot x^b = x^{a+b}$ $x^a/x^b = x^{a-b}$ $1/x^b = x^{-b}$ $(x^a)^b = x^{a.b}$ $(xy)^a = x^a \cdot y^a$	**Example:** $(x^2y)^3 = x^6y^3$

✎ **Simplify.**

1) $4^2 \cdot 4^2$ 4^4

2) $2 \cdot 2^2 \cdot 2^2$ 2^5

3) $3^2 \cdot 3^2$ 3

4) $3x^3 \cdot x$ $3x^4$

5) $12x^4 \cdot 3x$ $36x^5$

6) $6x \cdot 2x^2$ $12x^3$

7) $5x^4 \cdot 5x^4$ $25x^{16}$

8) $6x^2 \cdot 6x^3y^4$ $36x^5y^4$

9) $7x^2y^5 \cdot 9xy^3$ $63x^3y^8$

10) $7xy^4 \cdot 4x^3y^3$ $7x^4y^7$

11) $(2x^2)^2$ $2x^4$

12) $3x^5y^3 \cdot 8x^2y^3$ $3x^7y^6$

13) $7x^3 \cdot 10y^3x^5 \cdot 8yx^3$ $7x^4y^4$

14) $(x^4)^3$ x^{12}

15) $(2x^2)^4$ $2x^8$

16) $(x^2)^3$ x^6

17) $(6x)^2$ $6x^2$

18) $3x^4y^5 \cdot 7x^2y^3$ $21x^6y^8$

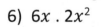
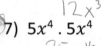
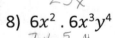

Division Property of Exponents

Helpful	$\frac{x^a}{x^b} = x^{a-b}$, $x \neq 0$	**Example:**
Hints		$\frac{x^{12}}{x^5} = x^7$

✎ **Simplify.**

1) $\frac{5^5}{5}$ 5^4

2) $\frac{3}{3^5}$ 3^{-4}

3) $\frac{2^2}{2^3}$ 2^{-1}

4) $\frac{2^4}{2^2}$ 2^2

5) $\frac{x}{x^3}$ x^{-2}

6) $\frac{3x^3}{9x^4}$ $\frac{1}{3}x^{-1}$

7) $\frac{2x^{-5}}{9x^{-2}}$ $\frac{2}{9}x^{-3}$

8) $\frac{21x^8}{7x^3}$ $3x^5$

9) $\frac{7x^6}{4x^7}$ $3x^{-1}$

10) $\frac{6x^2}{4x^3}$ $1.5x^{-1}$

11) $\frac{5x}{10x^3}$ $\frac{1}{2}x^{-2}$

12) $\frac{3x^3}{2x^5}$ $1.5x^{-2}$

13) $\frac{12x^3}{14x^6}$ $\frac{6}{7}x^{-3}$

14) $\frac{12x^3}{9y^8}$ $1\frac{1}{3}x^{-5}$

15) $\frac{25xy^4}{5x^6y^2}$ $5x^{-1}y^2$

16) $\frac{2x^4}{7x}$ $\frac{2}{7}x^3$

17) $\frac{16x^2y^8}{4x^3}$ $4x^{-1}y^8$

18) $\frac{12x^4}{15x^7y^9}$ $\frac{4}{5}x^{-3}y^9$

19) $\frac{12yx^4}{10yx^8}$ $1.2yx^{-4}$

20) $\frac{16x^4y}{9x^8y^2}$ $1\frac{7}{9}x^{-4}y^{-1}$

21) $\frac{5x^8}{20x^8}$ $\frac{1}{4}x^0$

Powers of Products and Quotients

Helpful	For any nonzero numbers a and b and any integer x, $(ab)^x = a^x \cdot b^x$.	**Example:**
Hints		$(2x^2 \cdot y^3)^2 =$
		$4x^2 \cdot y^6$

 Simplify.

1) $(2x^3)^4$

    $2x^{12}$

2) $(4xy^4)^2$

    $4x^2y^8$

3) $(5x^4)^2$

    $5x^8$

4) $(11x^5)^2$

    $11x^{10}$

5) $(4x^2y^4)^4$

    $4x^8y^{16}$

6) $(2x^4y^4)^3$

    $2x^{12}y^{12}$

7) $(3x^2y^2)^2$

    $3x^4y^4$

8) $(3x^4y^3)^4$

9) $(2x^6y^8)^2$

10) $(12x\ 3x)^3$

11) $(2x^9\ x^6)^3$

12) $(5x^{10}y^3)^3$

13) $(4x^3\ x^2)^2$

14) $(3x^3\ 5x)^2$

15) $(10x^{11}y^3)^2$

16) $(9x^7\ y^5)^2$

17) $(4x^4y^6)^5$

18) $(4x^4)^2$

19) $(3x\ 4y^3)^2$

20) $(9x^2y)^3$

21) $(12x^2y^5)^2$

Zero and Negative Exponents

Helpful *Hints*	A negative exponent simply means that the base is on the wrong side of the fraction line, so you need to flip the base to the other side. For instance, "x^{-2}" (pronounced as "ecks to the minus two") just means "x^2" but underneath, as in $\frac{1}{x^2}$	**Example:** $$5^{-2} = \frac{1}{25}$$

$\frac{49}{7}$
$\frac{6}{343}$

✎ *Evaluate the following expressions.*

1) 8^{-2} $\frac{1}{8^{-2}}$ $\frac{1}{64}$

2) 2^{-4} $\frac{1}{16}$

3) 10^{-2} $\frac{1}{100}$ $\frac{25}{5}$ $\frac{25}{125}$

4) 5^{-3} $\frac{1}{125}$

5) 22^{-1} $\frac{1}{22}$

6) 9^{-1} $\frac{1}{9}$

7) 3^{-2} $\frac{1}{27}$

8) 4^{-2} $\frac{1}{16}$

9) 5^{-2} $\frac{1}{25}$

10) 35^{-1} $\frac{1}{35}$ $\frac{36}{6}$ $\frac{6}{192}$

11) 6^{-3} $\frac{1}{192}$ $\frac{36}{192}$

12) 0^{15} 0

13) 10^{-9} $\frac{1}{1000000000}$

14) 3^{-4} $\frac{1}{81}$

15) 5^{-2} $\frac{1}{25}$

16) 2^{-3} $\frac{1}{8}$

17) 3^{-3} $\frac{1}{27}$

18) 8^{-1} $\frac{1}{8}$

19) 7^{-3} $\frac{1}{343}$

20) 6^{-2} $\frac{1}{36}$

21) $(\frac{2}{3})^{-2}$

22) $(\frac{1}{5})^{-3}$

23) $(\frac{1}{2})^{-8}$

24) $(\frac{2}{5})^{-3}$

25) 10^{-3} $\frac{1}{1000}$

26) 1^{-10} -1

Negative Exponents and Negative Bases

Helpful *Hints*	– Make the power positive. A negative exponent is the reciprocal of that number with a positive exponent. – The parenthesis is important! -5^{-2} is not the same as $(-5)^{-2}$ $-5^{-2} = -\frac{1}{5^2}$ and $(-5)^{-2} = +\frac{1}{5^2}$	**Example:** $2x^{-3} = \frac{2}{x^3}$

✏ *Simplify.*

1) -6^{-1} $-\frac{1}{6}$

2) $-4x^{-3}$ $-\frac{1}{4x^3}$

3) $-\frac{5x}{x^{-3}}$

4) $-\frac{a^{-3}}{b^{-2}}$

5) $-\frac{5}{x^{-3}}$

6) $\frac{7b}{-9c^{-4}}$

7) $-\frac{5n^{-2}}{10p^{-3}}$

8) $\frac{4ab^{-2}}{-3c^{-2}}$

9) $-12x^2y^{-3}$

10) $\left(-\frac{1}{3}\right)^{-2}$

11) $\left(-\frac{3}{4}\right)^{-2}$

12) $\left(\frac{3a}{2c}\right)^{-2}$

13) $\left(-\frac{5x}{3yz}\right)^{-3}$

14) $-\frac{2x}{a^{-4}}$

Writing Scientific Notation

Helpful

Hints

– It is used to write very big or very small numbers in decimal form.

– In scientific notation all numbers are written in the form of:

$$m \times 10^n$$

Decimal notation	Scientific notation
5	5×10^0
−25,000	-2.5×10^4
0.5	5×10^{-1}
2,122.456	$2,122456 \times 10^3$

✎ *Write each number in scientific notation.*

1) 91×10^3

9.1×10^4

2) 60

6×10^1

3) 2000000

2×10^6

4) 0.0000006

6×10^{-6}

5) 354000

3.54×10^5

6) 0.000325

3.25×10^{-4}

7) 2.5

2.5×10^0

8) 0.00023

2.3×10^{-4}

9) 56000000

5.6×10^7

10) 2000000

2×10^6

11) 78000000

7.8×10^7

12) 0.0000022

2.2×10^{-6}

13) 0.00012

1.2×10^{-4}

14) 0.004

4×10^{-3}

15) 78

7.8×10^1

16) 1600

1.6×10^3

17) 1450

1.45×10^3

18) 130000

1.3×10^5

19) 60

6×10^1

20) 0.113

1.13×10^{-1}

21) 0.02

2×10^{-2}

Square Roots

Helpful	— A square root of x is a number r whose square is: $r^2 = x$ **Example:**
Hints	r is a square root of x. $\sqrt{4} = 2$

📝 *Find the value each square root.*

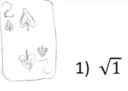

1) $\sqrt{1}$

1

2) $\sqrt{4}$

2

3) $\sqrt{9}$

3

4) $\sqrt{25}$

5

5) $\sqrt{16}$

4

6) $\sqrt{49}$

7

7) $\sqrt{36}$

6

8) $\sqrt{0}$

0

9) $\sqrt{64}$

8

10) $\sqrt{81}$

9

11) $\sqrt{121}$

11

12) $\sqrt{225}$

15

13) $\sqrt{144}$

12

14) $\sqrt{100}$

10

15) $\sqrt{256}$

16) $\sqrt{289}$

17) $\sqrt{324}$

18) $\sqrt{400}$

20

19) $\sqrt{900}$

30

20) $\sqrt{529}$

21) $\sqrt{90}$

Test Preparation

1) What is the value of 5^4 ?

   Write your answer in the box below.

675

2) How is this number written in scientific notation?

   0.00002389

   A. 2.389×10^{-5}

   B. 23.89×10^{6}

   C. 0.2389×10^{-4}

   D. 2389×10^{-8}

3) What is the value of 3^6?

   Write your answer in the box below.

81

4) How is this number written in scientific notation?

$$0.0050468$$

A. 5.0468×10^{-3}

B. 5.0468×10^{3}

C. 0.50468×10^{-2}

D. 50468×10^{-7}

Answers of Worksheets – Chapter 8

Multiplication Property of Exponents

1) 4^4
2) 2^5
3) 3^4
4) $3x^4$
5) $36x^5$
6) $12x^3$

7) $25x^8$
8) $36x^5y^4$
9) $63x^3y^8$
10) $28x^4y^7$
11) $4x^4$
12) $24x^7y^6$

13) $560x^{11}y^4$
14) x^{12}
15) $16x^8$
16) x^6
17) $36x^2$
18) $21x^6y^8$

Division Property of Exponents

1) 5^4
2) $\frac{1}{3^4}$
3) $\frac{1}{2}$
4) 2^2
5) $\frac{1}{x^2}$
6) $\frac{1}{3x}$
7) $\frac{2}{9x^3}$
8) $3x^5$

9) $\frac{7}{4x}$
10) $\frac{3}{2x}$
11) $\frac{1}{2x^2}$
12) $\frac{3}{2x^2}$
13) $\frac{6}{7x^3}$
14) $\frac{4x^3}{3y^8}$
15) $\frac{5y^2}{x^5}$

16) $\frac{2x^3}{7}$
17) $\frac{4y^8}{x}$
18) $\frac{4}{5x^3y^9}$
19) $\frac{6}{5x^4}$
20) $\frac{16}{9x^4y}$
21) $\frac{1}{4}$

Powers of Products and Quotients

1) $16x^{12}$
2) $16x^2y^8$
3) $25x^8$
4) $121x^{10}$
5) $256x^8y^{16}$
6) $8x^{12}y^{12}$

7) $9x^4y^4$
8) $81x^{16}y^{12}$
9) $4x^{12}y^{16}$
10) $46,656x^6$
11) $8x^{45}$
12) $125x^{30}y^9$

13) $16x^{10}$
14) $225x^8$
15) $100x^{22}y^6$
16) $81x^{14}y^{10}$
17) $1,024x^{20}y^{30}$
18) $16x^8$

19) $144x^2y^6$ 20) $729x^6y^3$ 21) $144x^4y^{10}$

Zero and Negative Exponents

1) $\frac{1}{64}$ 10) $\frac{1}{35}$ 19) $\frac{1}{343}$

2) $\frac{1}{16}$ 11) $\frac{1}{216}$ 20) $\frac{1}{36}$

3) $\frac{1}{100}$ 12) 0   21) $\frac{9}{4}$

4) $\frac{1}{125}$ 13) $\frac{1}{1000000000}$ 22) 125

5) $\frac{1}{22}$ 14) $\frac{1}{81}$ 23) 256

6) $\frac{1}{9}$ 15) $\frac{1}{25}$ 24) $\frac{125}{8}$

7) $\frac{1}{9}$ 16) $\frac{1}{8}$ 25) $\frac{1}{1000}$

8) $\frac{1}{16}$ 17) $\frac{1}{27}$ 26) 1

9) $\frac{1}{25}$ 18) $\frac{1}{8}$

Negative Exponents and Negative Bases

1) $-\frac{1}{6}$ 6) $-\frac{7bc^4}{9}$ 10) 9

2) $-\frac{4}{x^3}$ 7) $-\frac{p^3}{2n^2}$ 11) $\frac{16}{9}$

3) $-5x^4$ 8) $-\frac{4ac^2}{3b^2}$ 12) $\frac{4c^2}{9a^2}$

4) $-\frac{b^2}{a^3}$ 9) $-\frac{12x^2}{y^3}$ 13) $-\frac{27y^3z^3}{125x^3}$

5) $-5x^3$ 14) $-2xa^4$

Writing Scientific Notation

1) 9.1×10^4 4) 6×10^{-7} 7) 2.5×10^0

2) 6×10^1 5) 3.54×10^5 8) 2.3×10^{-4}

3) 2×10^6 6) 3.25×10^{-4} 9) 5.6×10^7

10) 2×10^6

11) 7.8×10^7

12) 2.2×10^{-6}

13) 1.2×10^{-4}

14) 4×10^{-3}

15) 7.8×10^1

16) 1.6×10^3

17) 1.45×10^3

18) 1.3×10^5

19) 6×10^1

20) 1.13×10^{-1}

21) 2×10^{-2}

Square Roots

1) 1

2) 2

3) 3

4) 5

5) 4

6) 7

7) 6

8) 0

9) 8

10) 9

11) 11

12) 15

13) 12

14) 10

15) 16

16) 17

17) 18

18) 20

19) 30

20) 23

21) $3\sqrt{10}$

Test Preparation Answers

1) The answer is 625.

$5^4 = 5 \times 5 \times 5 \times 5 = 625$

2) Choice A is correct.

$0.00002389 = \dfrac{2.389}{100000} \Rightarrow 2.389 \times 10^{-5}$

3) The answer is 729.

$3^6 = 3 \times 3 \times 3 \times 3 \times 3 \times 3 = 729$

4) Choice A is correct

$0.0050468 = \dfrac{5.0468}{1000} \Rightarrow 5.0468 \times 10^{-3}$

Chapter 9: Measurements

Topics that you'll learn in this chapter:

- ✓ Inches & Centimeters
- ✓ Metric units
- ✓ Distance Measurement
- ✓ Weight Measurement

Inches and Centimeters

Helpful *Hints*	1 inch = 2.5 cm	**Example:**
	1 foot = 12 inches	
	1 yard = 3 feet	18 inches = 0.4572 m
	1 yard = 36 inches	
	1 inch = 0.0254 m	

✍ **Convert to the units.**

1inch = 2.5 cm

1) 25 cm = _____ 10 inches

2) 11 inches = _____ cm

3) 1 m = _____ 275 inches

4) 80 inches = _____ 2.022 m

5) 200 cm = _____ 2.002 m

6) 5 m = _____ 500 cm

7) 4 feet = _____ 48 inches

8) 10 yards = _____ 360 inches

9) 16 feet = _____ 192 inches

10) 48 inches = _____ 4 Feet

11) 4 inches = _____ 10 cm

12) 12.5 cm = _____ 5 inches

13) 6 feet: _____ 72 inches

14) 10 feet: _____ 120 inches

15) 12 yards: _____ 432 feet

16) 7 yards: _____ 252 feet

132

Metric Units

Helpful	1 m = 100 cm	**Example:**
	1 cm = 10 mm	
Hints	1 m = 1000 mm	12 cm = 0.12 m
	1 km = 1000 m	

254
254
508

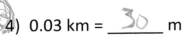Convert to the units.

1,000,000

1) 4 mm = __¼__ cm

2) 0.6 m = __600__ mm

3) 2 m = __200__ cm

4) 0.03 km = __30__ m

5) 3000 mm = __.003__ km

6) 5 cm = __.05__ m

7) 0.03 m = __30__ cm

8) 1000 mm = __.001__ km

9) 600 mm = __.6__ m

10) 0.77 km = __770,000__ mm

11) 0.08 km = __80__ m

12) 0.30 m = _____ cm

13) 400 m = _____ km

14) 5000 cm = _____ km

15) 40 mm = _____ cm

16) 800 m = _____ km

Distance Measurement

Helpful	1 mile = 5280 ft	Example:
	1 mile = 1760 yd	
Hints	1 mile = 1609.34 m	10 miles = 52800 ft

✎ **Convert to the new units.**

1) 2 mi = _____ ft

2) 21 mi = _____ ft

3) 6 mi = _____ ft

4) 3 mi = _____ yd

5) 72 mi = _____ ft

6) 41 mi = _____ yd

7) 62 mi = _____ yd

8) 39 mi = _____ yd

9) 7 mi = _____ yd

10) 94 mi = _____ yd

11) 87 mi = _____ yd

12) 23 mi = _____ yd

13) 2 mi = _____ m

14) 5 mi = _____ m

15) 6 mi = _____ m

16) 3 mi = _____ m

Weight Measurement

Helpful

Hints

1 kg = 1000g

Example:

2000 g = 2 kg

Convert to grams.

1) 0.5 kg = _____ g

2) 3.2 kg = _____ g

3) 8.2 kg = _____ g

4) 9.2 kg = _____ g

5) 35 kg = _____ g

6) 87 kg = _____ g

7) 45 kg = _____ g

8) 15 kg = _____ g

9) 0.32 kg = _____ g

10) 81 kg = _____ g

Convert to kilograms.

11) 200,000 g = _____ kg

12) 30,000 g = _____ kg

13) 800,000 g = _____ kg

14) 20,000 g = _____ kg

15) 40,000 g = _____ kg

16) 500,000 g = _____ kg

Test Preparation

1) 12 yards 4 feet and 2 inches equals to how many inches?

    A.   96

    B.   432

    C.   482

    D.   578

2) A rope weighs 800 grams per meter of length. What is the weight in kilograms of 12.2 meters of this rope? (1 kilograms = 1000 grams)

    A. 0.0976

    B. 0.976

    C. 9.76

    D. 9,760

3) A house floor has a perimeter of 4,221 feet. What is the perimeter of the floor in yards?

Write your answer in the box below.

Answers of Worksheets – Chapter 9

Inches & Centimeters

1) 25 cm = 9.84 inches

2) 11 inch = 27.94 cm

3) 1 m = 39.37 inches

4) 80 inch = 2.03 m

5) 200 cm = 2 m

6) 5 m = 500 cm

7) 4 feet = 48 inches

8) 10 yards = 360 inches

9) 16 feet = 192 inches

10) 48 inches = 4 Feet

11) 4 inch = 10.16 cm

12) 12.5 cm = 4.92 inches

13) 6 feet: 72 inches

14) 10 feet: 120 inches

15) 12 yards: 36 feet

16) 7 yards: 21 feet

Metric Units

1) 4 mm = 0.4 cm

2) 0.6 m = 600 mm

3) 2 m = 200 cm

4) 0.03 km = 30 m

5) 3000 mm = 0.003 km

6) 5 cm = 0.05 m

7) 0.03 m = 3 cm

8) 1000 mm = 0.001 km

9) 600 mm = 0.6 m

10) 0.77 km = 770,000 mm

11) 0.08 km = 80 m

12) 0.30 m = 30 cm

13) 400 m = 0.4 km

14) 5000 cm = 0.05 km

15) 40 mm = 4 cm

16) 800 m = 0.8 km

Distance Measurement

1) 21 mi = 110880 ft

2) 6 mi = 31680 ft

3) 3 mi = 5280 yd

4) 72 mi = 380160 ft

5) 41 mi = 72160 yd

6) 62 mi = 109120 yd

7) 39 mi = 68640 yd

8) 7 mi = 12320 yd

9) 94 mi = 165440 yd

10) 87 mi = 153120 yd

11) 23 mi = 40480 yd

12) 2 mi = 3218.69 m

13) 5 mi = 8046.72 m

14) 6 mi = 9656.06 m

15) 3 mi = 4828.03 m

Weight Measurement

1) 0.5 kg = 500 g

2) 3.2 kg = 3200 g

3) 8.2 kg = 8200 g

4) 9.2 kg = 9200 g

5) 35 kg = 35000 g

6) 87 kg = 87000 g

7) 45 kg = 45000 g

8) 15 kg = 15000 g

9) 0.32 kg = 320 g

10) 81 kg = 81000 g

11) 200,000 g = 200 kg

12) 30,000 g = 30 kg

13) 800,000 g = 800 kg

14) 20,000 g = 20 kg

15) 40,000 g = 40 kg

16) 500,000 g = 500 kg

Test Preparation Answers

1) Choice C is correct

12Yards = (12×36) 432 inches

4feet = (4×12) 48 inches

12 yards + 4 feet + 2 inches = 432 + 48 +2 = 482

2) Choice C is correct

The weighs of rope per meter of length = 800 grams. We use ratio to find answer (Let x be the amount of weight):

$$\frac{800 \text{ grams}}{1 \text{ meter}} = \frac{x \text{ grams}}{12.2 \text{ meter}} \Rightarrow x = 9760 \text{ grams} \quad x = 9.760 \text{ kilograms}$$

3) The answer is 1407.

1 yard=3feet

$4,221 \div 3 = 1,407$

Chapter 10: Plane Figures

Topics that you'll learn in this chapter:

- ✓ The Pythagorean Theorem
- ✓ Area of Triangles
- ✓ Perimeter of Polygons
- ✓ Area and Circumference of Circles
- ✓ Area of Squares, Rectangles, and Parallelograms
- ✓ Area of Trapezoids

The Pythagorean Theorem

Helpful

Hints

– In any right triangle:

$a^2 + b^2 = c^2$

Example:

Missing side = 6

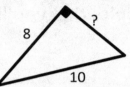

8

?

10

✍ *Do the following lengths form a right triangle?*

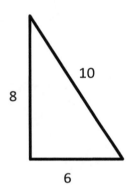

8   10   6

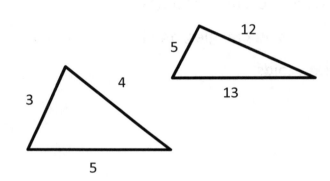

3   4   5

5   12   13

✍ *Find each missing length to the nearest tenth.*

4)

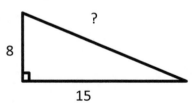

8   ?   15

5)

?   24   ?   10

6)

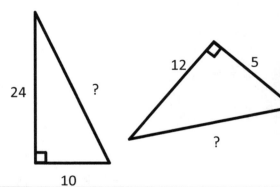

12   5   ?

Area of Triangles

Helpful   Area $=\frac{1}{2}$ ($base \times height$)

Hints

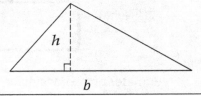

✎ *Find the area of each.*

1)

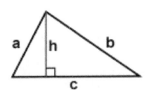

c = 9 mi

h = 3.7 mi

2)

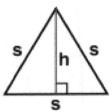

s = 14 m

h = 12.2 m

3)

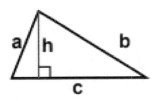

a = 5 m

b = 11 m

c = 14 m

h = 4 m

4)

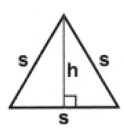

s = 10 m

h = 8.6 m

Perimeter of Polygons

Helpful

Hints

Perimeter of a square = 4s

 s

Perimeter of a rectangle

= 2(l + w)

w

l

Perimeter of trapezoid

= a + b + c + d

a
d   b
c

Perimeter of Pentagon = 6a

a

Perimeter of a parallelogram = 2(l + w)

l

w

Example:

P = 18

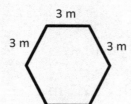

3 m

3 m   3 m

✎ *Find the perimeter of each shape.*

1)

5 m

5 m   5 m

2)

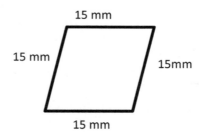

15 mm

15 mm   15mm

15 mm

3)

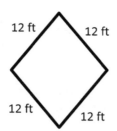

12 ft   12 ft

12 ft   12 ft

4)

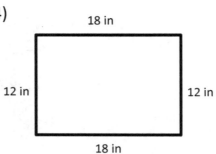

18 in

12 in   12 in

18 in

Area and Circumference of Circles

Helpful	Area = πr²	Example:

Circumference = 2πr

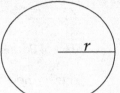

Hints

If the radius of a circle is 3, then:

Area = 28.27

Circumference = 18.85

✍ *Find the area and circumference of each.* (π = 3.14)

1)

4 in

2)

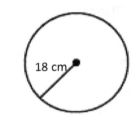

18 cm

3)

5 m

4)

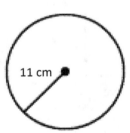

11 cm

5)

8 km

6)

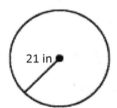

21 in

Area of Squares, Rectangles, and Parallelograms

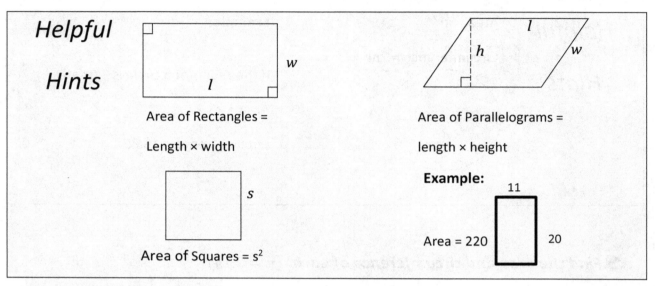

Helpful

Hints

Area of Rectangles =

Length × width

Area of Squares = s^2

Area of Parallelograms =

length × height

Example:

Area = 220

🖋 *Find the area of each.*

1)

22 yd

32.3 yd   32.3 yd

22 yd

2)

27mi

27 mi   27 mi

27 mi

3)

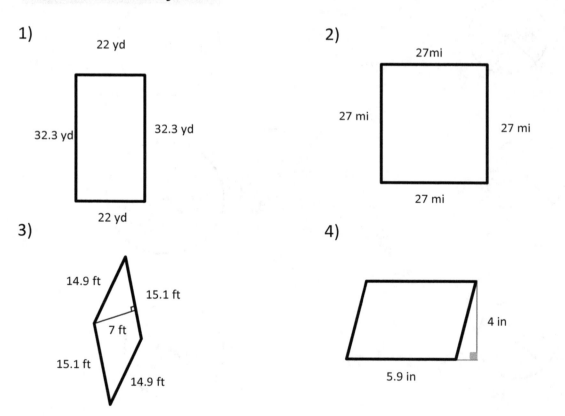

14.9 ft

15.1 ft

7 ft

15.1 ft

14.9 ft

4)

4 in

5.9 in

Area of Trapezoids

Helpful	$A = \frac{1}{2}h(b_1 + b_2)$	**Example:**
Hints		$A = 252 \text{ cm}^2$

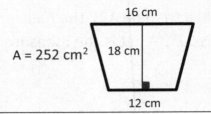

16 cm

18 cm

12 cm

✎ **Calculate the area for each trapezoid.**

1)

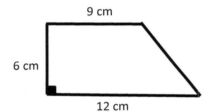

9 cm

6 cm

12 cm

2)

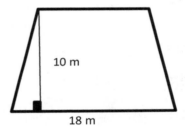

14 m

10 m

18 m

3)

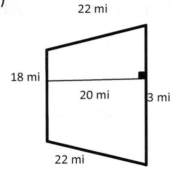

22 mi

18 mi

20 mi

3 mi

22 mi

4)

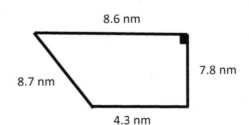

8.6 nm

8.7 nm

7.8 nm

4.3 nm

Test Preparation

1) In a triangle ABC the measure of angle ACB is 68° and the measure of angle CAB is 52°. What is the measure of angle ABC?

   Write your answer in the box below.

2) The perimeter of a rectangular yard is 60 meters. What is its length if its width is 10 meters?

   A. 10 meters

   B. 18 meters

   C. 20 meters

   D. 24 meters

3) The radius of the following cylinder is 4 inches and its height is 10 inches. What is the volume of the cylinder?

   Write your answer in the box below. (π equals 3.14) (Round your answer to the nearest whole number)

   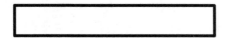

4) The perimeter of the trapezoid below is 52. What is its area?

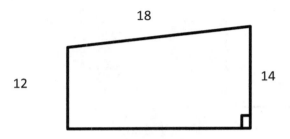

Write your answer in the box below.

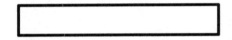

5) The area of a circle is 64 π. What is the circumference of the circle?

A. 8 π

B. 16 π

C. 32 π

D. 64 π

6) The length of a rectangle is 12 inches long and its area is 96 square inches. What is the perimeter of the rectangle?

Write your answer in the box below.

7) The perimeter of the trapezoid below is 36 cm. What is its area?

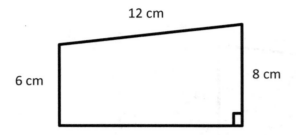

12 cm

6 cm

8 cm

A. 26

B. 42

C. 48

D. 70

8) What is the perimeter of a square that has an area of 169 square inches?
   Write your answer in the box below.

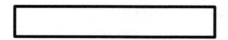

Answers of Worksheets – Chapter 10

The Pythagorean Theorem

1) yes
2) yes
3) yes

4) 17
5) 26
6) 13

Area of Triangles

1) 16.65 mi^2
2) 85.4 m^2

3) 28 m^2
4) 43 m^2

Perimeter of Polygons

1) 30 m
2) 60 mm

3) 48 ft
4) 60 in

Area and Circumference of Circles

1) Area: 50.24 in^2, Circumference: 25.12 in

2) Area: 1,017.36 cm^2, Circumference: 113.04 cm

3) Area: 78.5m^2, Circumference: 31.4 m

4) Area: 379.94 cm^2, Circumference: 69.08 cm

5) Area: 200.96 km^2, Circumference: 50.2 km

6) Area: 1,384.74 km^2, Circumference: 131.88 km

Area of Squares, Rectangles, and Parallelograms

1) 710.6 yd^2
2) 729 mi^2

3) 105.7 ft^2
4) 23.6 in^2

Area of Trapezoids

1) 63 cm^2
2) 160 m^2

3) 410 mi^2
4) 50.31 nm^2

Test Preparation Answers

1) The answer is 60.

The whole angles in every triangle are: 180°and Let x be the number of new angle so:

180 = 68 + 52 + $x \Rightarrow x = 60°$

2) Choice C is correct

Let x be the length, and its $width = 10$

Perimeter of the rectangle is 2(width + length) = $2(10 + x) = 60 \Rightarrow 10 + x = 30 \Rightarrow x = 20$

Length of the rectangle is 10 meters.

3) The answer is 502.

The volume of the cylinder: πr²h

The volume of the cylinder: $(3.14) \times (4)^2 \times 10 = 502.4 \cong 502$

4) The answer is 104.

The perimeter of the trapezoid = the sum of the lengths of its four sides

So: 52 = 12+ 18+ 14+ $x \Rightarrow x$ =8

The area of the trapezoid = the sum of its bases, multiply the half of its height

So The area of the trapezoid = $(12+14) \times \frac{8}{2}$ = 104

5) Choice B is correct

Use the formula of areas of circles.

Area = $\pi r^2 \Rightarrow 64\,\pi = \pi r^2 \Rightarrow 64 = r^2 \Rightarrow r = 8$

Radius of the circle is 8. Now, use the circumference formula:

Circumference = $2\pi r = 2\pi\,(8) = 16\,\pi$

6) The answer is 40.

Use the formula of areas of rectangles.

Area: length plus width $\Rightarrow 96 = 12 \times \text{width} \Rightarrow \text{width} = 8$

Use the formula of perimeter of rectangles.

Perimeter: 2(length + width) $\Rightarrow 2(12 + 8) = 40$

7) Choice D is correct

The perimeter of the trapezoid is 36 cm.

Therefore, the missing side (height) is $= 36 - 8 - 12 - 6 = 10$

Area of a trapezoid: $A = \frac{1}{2}\,h\,(b_1 + b_2) = \frac{1}{2}\,(10)\,(6 + 8) = 70$

8) The answer is 52.

Use the area of a square formula.

$S = a^2 \Rightarrow 169 = a^2 \Rightarrow a = 13$

Use the perimeter of a square formula.

$P = 4a \Rightarrow p = 4 \times 13 \Rightarrow p = 52$

Chapter 11: Solid Figures

Topics that you'll learn in this chapter:

- ✓ Volume of Cubes
- ✓ Volume of Rectangle Prisms
- ✓ Surface Area of Cubes
- ✓ Surface Area of Rectangle Prisms

Volume of Cubes

Helpful	− Volume is the measure of the amount of space inside of a solid figure, like a cube, ball, cylinder or pyramid.
Hints	− Volume of a cube = (one side)^3
	− Volume of a rectangle prism: Length × Width × Height

✎ *Find the volume of each.*

1)
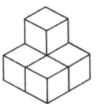

2)
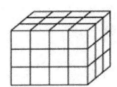

3)

4)

5)
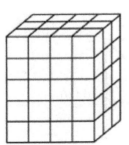

6)
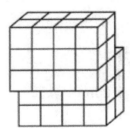

Volume of Rectangle Prisms

Helpful

Hints

Volume of rectangle prism

length × width × height

Example:

$10 \times 5 \times 8 = 400m^3$

10 m

8 m

5 m

✎ *Find the volume of each of the rectangular prisms.*

1)

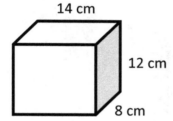

14 cm

12 cm

8 cm

2)

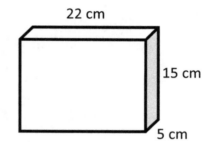

22 cm

15 cm

5 cm

3)

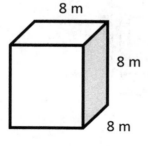

8 m

8 m

8 m

4)

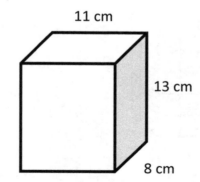

11 cm

13 cm

8 cm

Surface Area of Cubes

Helpful

Hints

Surface Area of a cube =

6 × (one side of the cube)²

Example:

$6 \times 4^2 = 96m^2$

4 m

4 m

4 m

✏️ **Find the surface of each cube.**

1)

6 mm

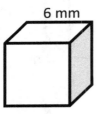

2)

9 mm

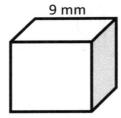

3)

10 cm

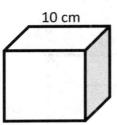

4)

8 m

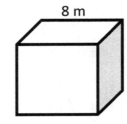

5)

7.5 in

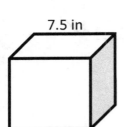

6)

11.3 ft

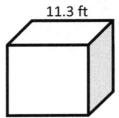

Surface Area of a Rectangle Prism

Helpful	
	Surface Area of a Rectangle Prism Formula:
Hints	SA =2 [(width × length) + (height × length) + width × height)]

✏️ **Find the surface of each prism.**

1)

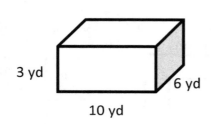

3 yd

6 yd

10 yd

2)

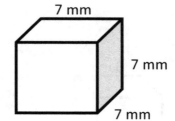

7 mm

7 mm

7 mm

3)

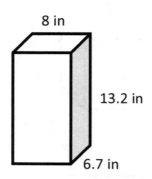

8 in

13.2 in

6.7 in

4)

17 cm

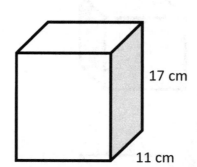

17 cm

11 cm

Volume of a Cylinder

✎ *Find the volume of each cylinder.* ($\pi = 3.14$)

1)

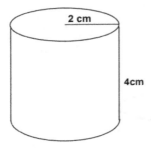

2)

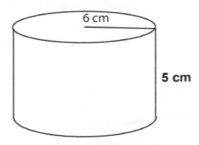

3)

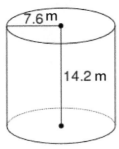

4)

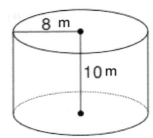

Surface Area of a Cylinder

Helpful

Hints

Surface area of a cylinder

SA = 2πr² + 2πrh

Example:

Surface area

= 1727

14 m

11 m

✎ **Find the surface of each cylinder.** (π = 3.14)

1)

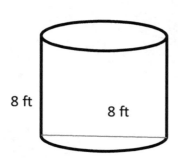

8 ft

8 ft

2)

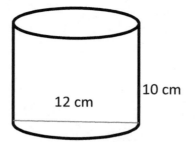

12 cm

10 cm

3)

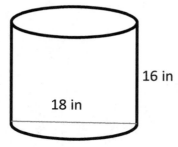

16 in

18 in

4)

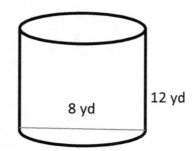

12 yd

8 yd

Test Preparation

1) A swimming pool holds 2,000 cubic feet of water. The swimming pool is 25 feet long and 10 feet wide. How deep is the swimming pool?

   Write your answer in the box below. (Don't write the measurement)

   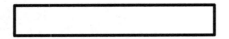

2) What is the volume of a cube whose side is 4 cm?

   A. 16 cm^3

   B. 32 cm^3

   C. 36 cm^3

   D. 64 cm^3

3) What is the volume of the cylinder below?

   A. $48 \pi \text{ in}^2$

   B. $57 \pi \text{ in}^2$

   C. $66 \pi \text{ in}^2$

   D. $72 \pi \text{ in}^2$

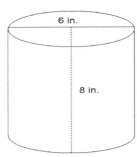

4) What is the volume of a box with the following dimensions?

Hight = 4 cm   Width = 5 cm   Length = 6 cm

   A. 15 cm^3
   B. 60 cm^3
   C. 90 cm^3
   D. 120 cm^3

Answers of Worksheets – Chapter 11

Volumes of Cubes

1) 8
2) 4
3) 5
4) 36
5) 60
6) 44

Volume of Rectangle Prisms

1) 1344 cm^3
2) 1650 cm^3
3) 512 m^3
4) 1144 cm^3

Surface Area of a Cube

1) 216 mm^2
2) 486 mm^2
3) 600 cm^2
4) 384 m^2
5) 337.5 in^2
6) 766.14 ft^2

Surface Area of a Prism

1) 216 yd^2
2) 294 mm^2
3) 495.28 in^2
4) 1326 cm^2

Volume of a Cylinder

1) 50.24 cm^3
2) 565.2 cm^3
3) 2,575.403 m^3
4) 2009.6 m^3

Surface Area of a Cylinder

1) 301.44 ft^2
2) 602.88 cm^2
3) 1413 in^2
4) 401.92 yd^2

Test Preparation Answers

1) The answer is 8.

Use formula of rectangle prism volume.

$V = (\text{length})\,(\text{width})\,(\text{height}) \Rightarrow 2000 = (25)\,(10)\,(\text{height}) \Rightarrow$

$\text{height} = 2000 \div 250 = 8$

2) Choice D is correct.

Use volume of a cube formula.

$V = a^3 \Rightarrow V = 4^3 \Rightarrow V = 64$

3) Choice D is correct.

Use volume of a cylinder formula.

$V = \pi\, r^2 h \Rightarrow V = \pi\, (3)^2 \text{ in} \times 8 \text{in} \Rightarrow V = 72\, \pi \text{ in}^2$

4) Choice D is correct

Volume of a box $= \text{length} \times \text{width} \times \text{height} = 4 \times 5 \times 6 = 120$

Chapter 12: Statistics

Topics that you'll learn in this chapter:

- ✓ Mean, Median, Mode, and Range of the Given Data
- ✓ The Pie Graph or Circle Graph
- ✓ Probability

Mean, Median, Mode, and Range of the Given Data

Helpful			**Example:**
	-	Mean: $\dfrac{\text{sum of the data}}{\text{of data entires}}$	
	-	Mode: value in the list that appears most often	22, 16, 12, 9, 7, 6, 4, 6
Hints	-	Range: largest value – smallest value	Mean = 10.25
			Mod = 6
			Range = 18

✏️ *Find Mean, Median, Mode, and Range of the Given Data.*

1) 7, 2, 5, 1, 1, 2

2) 2, 2, 2, 3, 6, 3, 7, 4

3) 9, 4, 3, 1, 7, 9, 4, 6, 4

4) 8, 4, 2, 4, 3, 2, 4, 5

5) 8, 5, 7, 5, 7, 9, 8

6) 5, 1, 4, 4, 9, 2, 9, 2, 5, 1

7) 4, 1, 5, 9, 7, 7, 5, 4, 3, 5

8) 7, 5, 4, 9, 6, 7, 7, 5, 2

9) 2, 5, 5, 6, 2, 4, 7, 6, 4, 9

10) 10, 5, 2, 5, 4, 5, 8, 10

11) 5, 1, 5, 2, 2

12) 2, 3, 5, 9, 6

The Pie Graph or Circle Graph

Helpful	A Pie Chart is a circle chart divided into sectors, each sector represents the relative size of each value.
Hints	

The circle graph below shows all Jason's expenses for last month. Jason spent $300 on his bills last month.

Answer following questions based on the Pie graph.

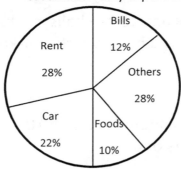

Jason's monthly expenses

1- How much did Jason spend on his car last month?

2- How much did Jason spend for foods last month?

3- How much did Jason spend on his rent last month?

4- What fraction is Jason's expenses for his bills and Car out of his total expenses last month?

5- How much was Jason's senses last month?

Probability of Simple Events

Helpful	-	Probability is the likelihood of something happening in the future. It is expressed as a number between zero (can never happen) to 1 (will always happen).	**Example:**
			Probability of a flipped coins turns up 'heads'
Hints	-	Probability can be expressed as a fraction, a decimal, or a percent.	Is $0.5 = \dfrac{1}{2}$

✎ *Solve.*

1) A number is chosen at random from 1 to 10. Find the probability of selecting a 4 or smaller.

2) There are 135 blue balls and 15 red balls in a basket. What is the probability of randomly choosing a red ball from the basket?

3) A number is chosen at random from 1 to 10. Find the probability of selecting of 4 and factors of 6.

4) What is the probability of choosing a Hearts in a deck of cards? (A deck of cards contains 52 cards)

5) A number is chosen at random from 1 to 50. Find the probability of selecting prime numbers.

6) A number is chosen at random from 1 to 25. Find the probability of not selecting a composite number.

Test Preparation

1) Anita's trick–or–treat bag contains 12 pieces of chocolate, 18 suckers, 18 pieces of gum, 24 pieces of licorice. If she randomly pulls a piece of candy from her bag, what is the probability of her pulling out a piece of sucker?

   A. $\dfrac{1}{3}$

   B. $\dfrac{1}{4}$

   C. $\dfrac{1}{6}$

2) The average of 6 numbers is 12. The average of 4 of those numbers is 10. What is the average of the other two numbers?

   A. 10

   B. 12

   C. 14

   D. 16

3) The average of 13, 15, 20 and x is 18. What is the value of x?

Write your answer in the box below.

4) What is the median of these numbers? 4, 9, 13, 8, 15, 18, 5
   A. 8

   B. 9

   C. 13

   D. 15

5) The mean of 50 test scores was calculated as 88. But, it turned out that one of the scores was misread as 94 but it was 69. What is the mean?
   A. 85

   B. 87

   C. 87.5

   D. 88.5

6) Two dice are thrown simultaneously, what is the probability of getting a sum of 6 or 9?

A. $\dfrac{1}{3}$

B. $\dfrac{1}{4}$

C. $\dfrac{1}{6}$

D. $\dfrac{1}{12}$

7) What is the median of these numbers? 2, 27, 28, 19, 67, 44, 35

A. 19

B. 28

C. 44

D. 35

8) The average of five numbers is 24. If a sixth number 42 is added, then, what is the new average?

    A. 25

    B. 26

    C. 27

    D. 28

Answers of Worksheets – Chapter 12

Mean, Median, Mode, and Range of the Given Data

1) mean: 3, median: 2, mode: 1, 2, range: 6
2) mean: 3.625, median: 3, mode: 2, range: 5
3) mean: 5.22, median: 4, mode: 4, range: 8
4) mean: 4, median: 4, mode: 4, range: 6
5) mean: 7, median: 7, mode: 5, 7, 8, range: 4
6) mean: 4.2, median: 4, mode: 1,2,4,5,9, range: 8
7) mean: 5, median: 5, mode: 5, range: 8
8) mean: 5.78, median: 6, mode: 7, range: 7
9) mean: 5, median: 5, mode: 2, 4, 5, 6, range: 7
10) mean: 6.125, median: 5, mode: 5, range: 8
11) mean: 3, median: 2, mode: 2, 5, range: 4
12) mean: 5, median: 5, mode: none, range: 7

The Pie Graph or Circle Graph

1) $550

2) $250

3) $700

4) $\frac{17}{50}$

5) $2500

Probability of simple events

1) $\frac{2}{5}$

2) $\frac{1}{10}$

3) $\frac{1}{5}$

4) $\frac{1}{4}$

5) $\frac{3}{10}$

6) $\frac{2}{5}$

Test Preparation Answers

1) Choice B is correct.

$$\text{Probability} = \frac{number\ of\ desired\ outcomes}{number\ of\ total\ outcomes} = \frac{18}{12+18+18+24} = \frac{18}{72} = \frac{1}{4}$$

2) Choice D is correct

average $= \frac{\text{sum of terms}}{\text{number of terms}} \Rightarrow$ (average of 6 numbers) $12 = \frac{\text{sum of numbers}}{6} \Rightarrow$ sum of 6 numbers is 12 × 6 = 72

(average of 4 numbers) $10 = \frac{\text{sum of numbers}}{4} \Rightarrow$ sum of 4 numbers is 10 × 4 = 40

sum of 6 numbers − sum of 4 numbers = sum of 2 numbers

72 − 40 = 32, average of 2 numbers $= \frac{32}{2}$ = 16

3) The answer is 24.

average $= \frac{\text{sum of terms}}{\text{number of terms}} \Rightarrow$ (average of 4 numbers) $18 = \frac{\text{sum of numbers}}{4} \Rightarrow$ sum of 4 numbers is : 13+ 15+ 20 + x

72 = sum of numbers $\Rightarrow$ 72= 13+ 15+ 20 + x $\Rightarrow x$=24

4) Choice B is correct

Write the numbers in order:

4, 5, 8, 9, 13, 15, 18

Since we have 7 numbers (7 is odd), then the median is the number in the middle, which is 9.

5) Choice C is correct

average (mean) $= \frac{\text{sum of terms}}{\text{number of terms}} \Rightarrow 88 = \frac{\text{sum of terms}}{50} \Rightarrow$ sum = 88 × 50 = 4,400

The difference of 94 and 69 is 25. Therefore, 25 should be subtracted from the sum.

4400 − 25 = 4375

mean = $\frac{\text{sum of terms}}{\text{number of terms}}$ ⇒ mean = $\frac{4375}{50}$ = 87.5

6) Choice B is correct

For Sum 6: (1 & 5) and (5 & 1), (2 & 4) and (4 & 2), (3 & 3), so we have 5 options

For sum 9: (3 & 6) and (6 & 3), (4 & 5) and (5 & 4), we have 4 options.

To get a sum of 6 or 9 for two dice: 7+4=9

Since, we have 6 × 6 = 36 total options, the probability of getting a sum of 6 and 9 is 11 out of 36 or $\frac{9}{36} = \frac{1}{4}$

7) Choice B is correct

Write the numbers in order:

2, 19, 27, 28, 35, 44, 67

Median is the number in the middle. So, the median is 28.

8) Choice C is correct

Solve for the sum of five numbers.

average = $\frac{\text{sum of terms}}{\text{number of terms}}$ ⇒ 24 = $\frac{\text{sum of 5 numbers}}{5}$ ⇒ sum of 5 numbers = 24 × 5 = 120

The sum of 5 numbers is 120. If a sixth number 42 is added, then the sum of 6 numbers is

120 + 42 = 162

average = $\frac{\text{sum of terms}}{\text{number of terms}}$ = $\frac{162}{6}$ = 27

ISEE Middle Level Practice Tests

The Independent School Entrance Exam (ISEE) is an admission test developed by the Educational Records Bureau for its member schools as part of their admission process.

ISEE Middle Level tests use a multiple-choice format and contain two Mathematics sections:

Quantitative Reasoning

There are 37 questions in the Quantitative Reasoning section and students have 35 minutes to answer the questions. This section contains word problems and quantitative comparisons. The word problems require either no calculation or simple calculation. The quantitative comparison items present two quantities, (A) and (B), and the student needs to select one of the following four answer choices:

(A) The quantity in Column A is greater.

(B) The quantity in Column B is greater.

(C) The two quantities are equal.

(D) The relationship cannot be determined from the information given.

Mathematics Achievement

There are 47 questions in the Mathematics Achievement section and students have 40 minutes to answer the questions. Mathematics Achievement measures students' knowledge of Mathematics requiring one or more steps in calculating the answer.

In this section, there are two complete ISEE Middle Level Quantitative Reasoning and Mathematics Achievement Tests. Let your student take these tests to see what score they'll be able to receive on a real ISEE test.

Good luck!

Time to Test

Time to refine your skill with a practice examination

Take a practice ISEE Middle Level Math Test to simulate the test day experience. After you've finished, score your test using the answer key.

Before You Start

- You'll need a pencil and scratch papers to take the test.

- For each question, there are four possible answers. Choose which one is best.

- It's okay to guess. You won't lose any points if you're wrong.

- Use the answer sheet provided to record your answers.

- After you've finished the test, review the answer key to see where you went wrong.

- **Calculators are NOT allowed for the ISEE Middle Level Test.**

Good Luck

ISEE Middle Level Practice Test Answer Sheets

Remove (or photocopy) these answer sheets and use them to complete the practice tests.

ISEE Middle Level Practice Test 1
Quantitative Reasoning

1) Ⓐ Ⓑ Ⓒ Ⓓ   2) Ⓐ Ⓑ Ⓒ Ⓓ

3) Ⓐ Ⓑ Ⓒ Ⓓ   4) Ⓐ Ⓑ Ⓒ Ⓓ

5) Ⓐ Ⓑ Ⓒ Ⓓ   6) Ⓐ Ⓑ Ⓒ Ⓓ

7) Ⓐ Ⓑ Ⓒ Ⓓ   8) Ⓐ Ⓑ Ⓒ Ⓓ

9) Ⓐ Ⓑ Ⓒ Ⓓ   10) Ⓐ Ⓑ Ⓒ Ⓓ

11) Ⓐ Ⓑ Ⓒ Ⓓ   12) Ⓐ Ⓑ Ⓒ Ⓓ

13) Ⓐ Ⓑ Ⓒ Ⓓ   14) Ⓐ Ⓑ Ⓒ Ⓓ

15) Ⓐ Ⓑ Ⓒ Ⓓ   16) Ⓐ Ⓑ Ⓒ Ⓓ

17) Ⓐ Ⓑ Ⓒ Ⓓ   18) Ⓐ Ⓑ Ⓒ Ⓓ

19) Ⓐ Ⓑ Ⓒ Ⓓ   20) Ⓐ Ⓑ Ⓒ Ⓓ

21) Ⓐ Ⓑ Ⓒ Ⓓ   22) Ⓐ Ⓑ Ⓒ Ⓓ

23) Ⓐ Ⓑ Ⓒ Ⓓ   24) Ⓐ Ⓑ Ⓒ Ⓓ

25) Ⓐ Ⓑ Ⓒ Ⓓ   26) Ⓐ Ⓑ Ⓒ Ⓓ

27) Ⓐ Ⓑ Ⓒ Ⓓ   28) Ⓐ Ⓑ Ⓒ Ⓓ

29) Ⓐ Ⓑ Ⓒ Ⓓ   30) Ⓐ Ⓑ Ⓒ Ⓓ

31) Ⓐ Ⓑ Ⓒ Ⓓ   32) Ⓐ Ⓑ Ⓒ Ⓓ

33) Ⓐ Ⓑ Ⓒ Ⓓ   34) Ⓐ Ⓑ Ⓒ Ⓓ

35) Ⓐ Ⓑ Ⓒ Ⓓ   36) Ⓐ Ⓑ Ⓒ Ⓓ

37) Ⓐ Ⓑ Ⓒ Ⓓ

ISEE Middle Level Practice Test 1
Mathematics Achievement

1) Ⓐ Ⓑ Ⓒ Ⓓ   2) Ⓐ Ⓑ Ⓒ Ⓓ
3) Ⓐ Ⓑ Ⓒ Ⓓ   4) Ⓐ Ⓑ Ⓒ Ⓓ
5) Ⓐ Ⓑ Ⓒ Ⓓ   6) Ⓐ Ⓑ Ⓒ Ⓓ
7) Ⓐ Ⓑ Ⓒ Ⓓ   8) Ⓐ Ⓑ Ⓒ Ⓓ
9) Ⓐ Ⓑ Ⓒ Ⓓ   10) Ⓐ Ⓑ Ⓒ Ⓓ
11) Ⓐ Ⓑ Ⓒ Ⓓ   12) Ⓐ Ⓑ Ⓒ Ⓓ
13) Ⓐ Ⓑ Ⓒ Ⓓ   14) Ⓐ Ⓑ Ⓒ Ⓓ
15) Ⓐ Ⓑ Ⓒ Ⓓ   16) Ⓐ Ⓑ Ⓒ Ⓓ
17) Ⓐ Ⓑ Ⓒ Ⓓ   18) Ⓐ Ⓑ Ⓒ Ⓓ
19) Ⓐ Ⓑ Ⓒ Ⓓ   20) Ⓐ Ⓑ Ⓒ Ⓓ
21) Ⓐ Ⓑ Ⓒ Ⓓ   22) Ⓐ Ⓑ Ⓒ Ⓓ
23) Ⓐ Ⓑ Ⓒ Ⓓ   24) Ⓐ Ⓑ Ⓒ Ⓓ
25) Ⓐ Ⓑ Ⓒ Ⓓ   26) Ⓐ Ⓑ Ⓒ Ⓓ
27) Ⓐ Ⓑ Ⓒ Ⓓ   28) Ⓐ Ⓑ Ⓒ Ⓓ
29) Ⓐ Ⓑ Ⓒ Ⓓ   30) Ⓐ Ⓑ Ⓒ Ⓓ
31) Ⓐ Ⓑ Ⓒ Ⓓ   32) Ⓐ Ⓑ Ⓒ Ⓓ
33) Ⓐ Ⓑ Ⓒ Ⓓ   34) Ⓐ Ⓑ Ⓒ Ⓓ
35) Ⓐ Ⓑ Ⓒ Ⓓ   36) Ⓐ Ⓑ Ⓒ Ⓓ
37) Ⓐ Ⓑ Ⓒ Ⓓ   38) Ⓐ Ⓑ Ⓒ Ⓓ
39) Ⓐ Ⓑ Ⓒ Ⓓ   40) Ⓐ Ⓑ Ⓒ Ⓓ
41) Ⓐ Ⓑ Ⓒ Ⓓ   42) Ⓐ Ⓑ Ⓒ Ⓓ
43) Ⓐ Ⓑ Ⓒ Ⓓ   44) Ⓐ Ⓑ Ⓒ Ⓓ
45) Ⓐ Ⓑ Ⓒ Ⓓ   46) Ⓐ Ⓑ Ⓒ Ⓓ
47) Ⓐ Ⓑ Ⓒ Ⓓ

ISEE Middle Level Practice Test 2
Quantitative Reasoning

1) Ⓐ Ⓑ Ⓒ Ⓓ   2) Ⓐ Ⓑ Ⓒ Ⓓ

3) Ⓐ Ⓑ Ⓒ Ⓓ   4) Ⓐ Ⓑ Ⓒ Ⓓ

5) Ⓐ Ⓑ Ⓒ Ⓓ   6) Ⓐ Ⓑ Ⓒ Ⓓ

7) Ⓐ Ⓑ Ⓒ Ⓓ   8) Ⓐ Ⓑ Ⓒ Ⓓ

9) Ⓐ Ⓑ Ⓒ Ⓓ   10) Ⓐ Ⓑ Ⓒ Ⓓ

11) Ⓐ Ⓑ Ⓒ Ⓓ   12) Ⓐ Ⓑ Ⓒ Ⓓ

13) Ⓐ Ⓑ Ⓒ Ⓓ   14) Ⓐ Ⓑ Ⓒ Ⓓ

15) Ⓐ Ⓑ Ⓒ Ⓓ   16) Ⓐ Ⓑ Ⓒ Ⓓ

17) Ⓐ Ⓑ Ⓒ Ⓓ   18) Ⓐ Ⓑ Ⓒ Ⓓ

19) Ⓐ Ⓑ Ⓒ Ⓓ   20) Ⓐ Ⓑ Ⓒ Ⓓ

21) Ⓐ Ⓑ Ⓒ Ⓓ   22) Ⓐ Ⓑ Ⓒ Ⓓ

23) Ⓐ Ⓑ Ⓒ Ⓓ   24) Ⓐ Ⓑ Ⓒ Ⓓ

25) Ⓐ Ⓑ Ⓒ Ⓓ   26) Ⓐ Ⓑ Ⓒ Ⓓ

27) Ⓐ Ⓑ Ⓒ Ⓓ   28) Ⓐ Ⓑ Ⓒ Ⓓ

29) Ⓐ Ⓑ Ⓒ Ⓓ   30) Ⓐ Ⓑ Ⓒ Ⓓ

31) Ⓐ Ⓑ Ⓒ Ⓓ   32) Ⓐ Ⓑ Ⓒ Ⓓ

33) Ⓐ Ⓑ Ⓒ Ⓓ   34) Ⓐ Ⓑ Ⓒ Ⓓ

35) Ⓐ Ⓑ Ⓒ Ⓓ   36) Ⓐ Ⓑ Ⓒ Ⓓ

37) Ⓐ Ⓑ Ⓒ Ⓓ

ISEE Middle Level Practice Test 2
Mathematics Achievement

1) Ⓐ Ⓑ Ⓒ Ⓓ   2) Ⓐ Ⓑ Ⓒ Ⓓ

3) Ⓐ Ⓑ Ⓒ Ⓓ   4) Ⓐ Ⓑ Ⓒ Ⓓ

5) Ⓐ Ⓑ Ⓒ Ⓓ   6) Ⓐ Ⓑ Ⓒ Ⓓ

7) Ⓐ Ⓑ Ⓒ Ⓓ   8) Ⓐ Ⓑ Ⓒ Ⓓ

9) Ⓐ Ⓑ Ⓒ Ⓓ   10) Ⓐ Ⓑ Ⓒ Ⓓ

11) Ⓐ Ⓑ Ⓒ Ⓓ   12) Ⓐ Ⓑ Ⓒ Ⓓ

13) Ⓐ Ⓑ Ⓒ Ⓓ   14) Ⓐ Ⓑ Ⓒ Ⓓ

15) Ⓐ Ⓑ Ⓒ Ⓓ   16) Ⓐ Ⓑ Ⓒ Ⓓ

17) Ⓐ Ⓑ Ⓒ Ⓓ   18) Ⓐ Ⓑ Ⓒ Ⓓ

19) Ⓐ Ⓑ Ⓒ Ⓓ   20) Ⓐ Ⓑ Ⓒ Ⓓ

21) Ⓐ Ⓑ Ⓒ Ⓓ   22) Ⓐ Ⓑ Ⓒ Ⓓ

23) Ⓐ Ⓑ Ⓒ Ⓓ   24) Ⓐ Ⓑ Ⓒ Ⓓ

25) Ⓐ Ⓑ Ⓒ Ⓓ   26) Ⓐ Ⓑ Ⓒ Ⓓ

27) Ⓐ Ⓑ Ⓒ Ⓓ   28) Ⓐ Ⓑ Ⓒ Ⓓ

29) Ⓐ Ⓑ Ⓒ Ⓓ   30) Ⓐ Ⓑ Ⓒ Ⓓ

31) Ⓐ Ⓑ Ⓒ Ⓓ   32) Ⓐ Ⓑ Ⓒ Ⓓ

33) Ⓐ Ⓑ Ⓒ Ⓓ   34) Ⓐ Ⓑ Ⓒ Ⓓ

35) Ⓐ Ⓑ Ⓒ Ⓓ   36) Ⓐ Ⓑ Ⓒ Ⓓ

37) Ⓐ Ⓑ Ⓒ Ⓓ   38) Ⓐ Ⓑ Ⓒ Ⓓ

39) Ⓐ Ⓑ Ⓒ Ⓓ   40) Ⓐ Ⓑ Ⓒ Ⓓ

41) Ⓐ Ⓑ Ⓒ Ⓓ   42) Ⓐ Ⓑ Ⓒ Ⓓ

43) Ⓐ Ⓑ Ⓒ Ⓓ   44) Ⓐ Ⓑ Ⓒ Ⓓ

45) Ⓐ Ⓑ Ⓒ Ⓓ   46) Ⓐ Ⓑ Ⓒ Ⓓ

47) Ⓐ Ⓑ Ⓒ Ⓓ

ISEE Middle Level

Practice Test 1

Quantitative Reasoning

- o **37 questions**
- o **Total time for this section:** 35 Minutes
- o **Calculators are not allowed at the test.**

1) Which of the following shows the numbers in descending order?

A. $\frac{1}{6}, \frac{2}{5}, \frac{1}{3}, \frac{3}{4}$

B. $\frac{1}{6}, \frac{2}{5}, \frac{3}{4}, \frac{1}{3}$

C. $\frac{1}{6}, \frac{1}{3}, \frac{2}{5}, \frac{3}{4}$

D. $\frac{1}{6}, \frac{3}{4}, \frac{1}{3}, \frac{2}{5}$

2) $452,357,841 \times 0.0001$?

A. 452,357.841

B. 45,235.7841

C. 4,523.57841

D. 452.357841

3) Jim purchased a table for 20% off and saved $23. What was the original price of the table?

A. $90

B. $100

C. $115

D. 140

4) A $40 shirt now selling for $28 is discounted by what percent?

    A. 20 %

    B. 30 %

    C. 40 %

    D. 60 %

5) If $f = 2x - 3y$ and $g = x + 4y$, what is $2f + g$?

    A. $3x - y$

    B. $3x - 2y$

    C. $5x - 2y$

    D. $5x - y$

6) Solve. $\dfrac{-48 \times 0.5}{6}$

    A. -16

    B. -4

    C. 4

    D. 16

7) What is the value of x in the following equation? $8^x = 512$

    A. 2
    B. 3
    C. 4
    D. 5

8) The score of Emma was half as that of Ava and the score of Mia was twice that of Ava. If the score of Mia was 60, what is the score of Emma?

    A. 15
    B. 18
    C. 20
    D. 30

9) The area of a circle is 64 π. What is the circumference of the circle?

    A. 8 π
    B. 16 π
    C. 32 π
    D. 64 π

10) Round off the result of 1.12×7.2 to the nearest tenth?

A. 8

B. 7

C. 8.06

D. 8.1

11) What is the value of x in the following figure?

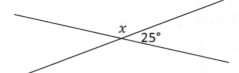

A. 25°

B. 65°

C. 115°

D. 155°

12) Two third of 18 is equal to $\frac{2}{5}$ of what number?

A. 60

B. 20

C. 12

D. 30

13) In five successive hours, a car travels 40 km, 45 km, 50 km, 35 km and 55 km. In the next five hours, it travels with an average speed of 50 km per hour. Find the total distance the car traveled in 10 hours.

    A. 425 km

    B. 450 km

    C. 475 km

    D. 500 km

14) What is the mean in the following set of numbers?

$$9, 12, 29, 36, 45, 63, 99, 123$$

    A. 46.2

    B. 40.5

    C. 59.4

    D. 52

15) The perimeter of the trapezoid below is 54. What is its area?

    A. 252 cm^2

    B. 234 cm^2

    C. 216 cm^2

    D. 130 cm^2

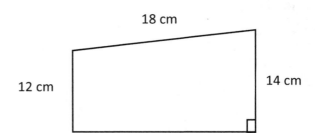

18 cm

12 cm

14 cm

16) The price of a laptop is decreased by 10% to $360. What is its original price?

    A. $320

    B. $380

    C. $390

    D. $400

17) Find $\frac{1}{4}$ of $\frac{2}{5}$ of 120?

    A. 16

    B. 12

    C. 8

    D. 4

18) The ratio of boys and girls in a class is 4:7. If there are 44 students in the class, how many more boys should be enrolled to make the ratio 1:1?

    A. 8

    B. 10

    C. 12

    D. 16

19) What is the value of x in the following equation?

$$4(x + 1) = 6(x - 4) + 20$$

A. 12

B. 4

C. 6.2

D. 5.5

20) A company pays its employee $7,000 plus 2% of all sales profit. If x is all sold profit, which of the following represents the employee's revenue?

A. $0.02x$

B. $0.98x - 7,000$

C. $0.02x + 7,000$

D. $0.98x + 7,000$

21) Which of the following is a correct statement?

A. $\frac{3}{4} > 0.8$

B. $10\% = \frac{2}{5}$

C. $3 < \frac{5}{2}$

D. $\frac{5}{6} > 0.8$

22) What is the value of x in the following figure?

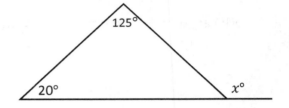

    A. 150

    B. 145

    C. 125

    D. 105

23) What is the area of a square whose diagonal is 8?

    A. 16

    B. 32

    C. 36

    D. 64

24) What is value of $-26 - (-65)$?

    A. 39

    B. -91

    C. -39

    D. 91

25) Car A use 8-liter petrol per 100 kilometers; car B use 6-liter petrol per 100 kilometers. If both cars drive 250 kilometers, how much more petrol does car A use?

    A. 5
    B. 10
    C. 15
    D. 20

Quantitative Comparisons

Direction: Questions 26 to 37 are Quantitative Comparisons Questions. Using the information provided in each question, compare the quantity in column A to the quantity in Column B. Choose on your answer sheet grid

 A if the quantity in Column A is greater

 B if the quantity in Column B is greater

 C if the two quantities are equal

 D if the relationship cannot be determined from the information given

26)

Column A	**Column B**
$6 + 4 \times 7 + 8$	$4 + 6 \times 7 - 8$

27) $y = -4x - 8$

Column A	**Column B**
The value of x when $y = 12$	-4

28)

Column A	**Column B**
$\sqrt{36} + \sqrt{36}$	$\sqrt{72}$

29) The average age of Joe, Michelle, and Nicole is 32.

Column A	**Column B**
The average age of Joe and Michelle	The average age of Michelle and Nicole

30)

Column A	**Column B**
$\sqrt{121 - 64}$	$\sqrt{121} - \sqrt{64}$

31) A right cylinder with radius 2 inches has volume 50π cubic inches.

Quantity A	Quantity B
The height of the cylinder	10 inches

32) x is an integer greater than zero.

Quantity A	Quantity B
$\frac{1}{x} + x$	8

33)

$$\frac{4}{5} < x < \frac{6}{7}$$

Quantity A	Quantity B
x	$\frac{5}{6}$

34)

Quantity A	Quantity B
$\frac{x^6}{6}$	$\left(\frac{x}{6}\right)^6$

35) The average of 3, 4, and x is 3.

Quantity A	Quantity B
x	average of $x, x - 6, x + 4, 2x$

36) a and b are real numbers.

$$a < b$$

Quantity A	Quantity B
$|a - b|$	$|b - a|$

37) $2x^3 + 10 = 64$

$\quad 120 - 18y = 84$

Quantity A	Quantity B
x	y

ISEE Middle Level

Practice Test 1

Mathematics Achievement

- o **47 questions**
- o **Total time for this section:** 40 Minutes
- o **Calculators are not allowed at the test.**

1) What number is 12 more than 15% of 180?

A. 40

B. 39

C. 15

D. 10

2) If a box contains red and blue balls in ratio of 2 : 3, how many red balls are there if 90 blue balls are in the box?

A. 90

B. 60

C. 30

D. 10

$$1 - \frac{1}{2} + 4$$

3) $3\left(\frac{1}{3} - \frac{1}{6}\right) + 4$?

A. 4

B. 4.5

C. 5.5

D. 5

4) In a bundle of 90 pencils, 43 are red and the rest are blue. About what percent of the bundle is composed of blue pencils?

    A. 62%

    B. 58%

    C. 54%

    D. 52%

5) What is the value of x in the following equation?
$$(x + 5)^3 = 64$$

    A. 1

    B. -1

    C. 2

    D. -2

6) What number is 5 less than 50% of 36?

    A. 12

    B. 13

    C. 18

    D. 23

7) What is the difference in perimeter between a 7 cm by 4 cm rectangle and a circle with diameter of 10 cm? ($\pi = 3$)

    A. 8 cm

    B. 9 cm

    C. 10 cm

    D. 11 cm

8) The price of a car was $20,000 in 2014, $16,000 in 2015 and $12,800 in 2016. What is the rate of depreciation of the price of car per year?

    A. 15 %

    B. 20 %

    C. 25 %

    D. 30 %

9) When a number is subtracted from 24 and the difference is divided by that number, the result is 3. What is the value of the number?

    A. 2

    B. 4

    C. 6

    D. 12

10) 90 is equal to?

    A. $2 + (3 \times 10) + (2 \times 30)$

    B. $\left(\frac{10}{3} \times 27\right) + (\frac{5}{2} \times 2)$

    C. $\left(\left(\frac{3}{2} + 3\right) \times \frac{18}{3}\right) + 63$

    D. $(2 \times 15) + (50 \times 2) - 46$

11) If $\frac{3x}{2} = 30$, then $\frac{2x}{5} = ?$

    A. 8

    B. 10

    C. 15

    D. 20

12) Which of the following is the greatest number?

    A. $\frac{1}{2}$

    B. $\frac{7}{9}$

    C. 0.9

    D. 65%

13) Calculate the approximate area of the following circle.

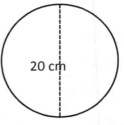

A. 1,257 cm²

B. 314 cm²

C. 126 cm²

D. 63 cm²

14) If 150 % of a number is 75, then what is the 90 % of that number?

A. 45

B. 50

C. 70

D. 85

15) What is the missing term in the given numbers?

2, 3, 5, 8, 12, 17, 23, ___, 38

A. 24

B. 26

C. 27

D. 30

16) Which of the following angles can represent the three angles of an isosceles right triangle?

    A. 10°, 80°, 90°

    B. 50°, 50°, 80°

    C. 60°, 60°, 60°

    D. 45°, 45°, 90°

17) In following rectangle which statement is true?

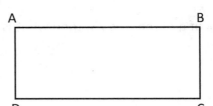

    A. AB is parallel to BC.

    B. The sum of all the angles equals 180°.

    C. Length of AB equal to length DC.

    D. AB is perpendicular to DC.

18) When a gas tank can hold 25 gallons, how many gallons does it contain when it is $\frac{2}{5}$ full?

    A. 125

    B. 62.5

    C. 50

    D. 10

19) A football team had $20,000 to spend on supplies. The team spent $14,000 on new balls. New sport shoes cost $120 each. Which of the following inequalities represent how many new shoes the team can purchase?

    A. $120x + 14,000 \leq 20,000$

    B. $120x + 14,000 \geq 20,000$

    C. $14,000x + 120 \leq 20,000$

    D. $14,000x + 12,0 \geq 20,000$

20) The capacity of a red box is 20% greater than a blue box. If the capacity of the red box is 30 books, how many books can be put in the blue box?

    A. 9

    B. 15

    C. 25

    D. 30

21) From last year, the price of gasoline has increased from $1.25 per gallon to $1.75 per gallon. The new price is what percent of the original price?

    A. 72 %

    B. 120 %

    C. 140 %

    D. 160 %

22) 195 minutes = ...?

    A. 3.25 Hours

    B. 2.16 Hours

    C. 2 Hours

    D. 0.3 Hours

23) If a gas tank can hold 25 gallons, how many gallons does it contain when it is $\frac{2}{5}$ full?

    A. 50

    B. 125

    C. 62.5

    D. 10

24) Which of the following is **not** a prime number?

    A. 103

    B. 101

    C. 97

    D. 57

25) What is the perimeter of a square that has an area of 64 square inches?

A. 144 inches

B. 64 inches

C. 32 inches

D. 56 inches

26) Jason left a $12.00 tip on a lunch that cost $48.00, approximately what percentage was the tip?

A. 2.5%

B. 10%

C. 15%

D. 25%

27) Two-kilograms apple and three-kilograms orange cost $26.4 If one-kilogram apple costs $4.2 how much does one-kilogram orange cost?

A. $9

B. $6

C. $5.5

D. $5

28) $\left(\left((-12) + 20\right) \times 2\right) + (-15)$?

    A. 1

    B. 2

    C. 3

    D. 4

29) The width of a rectangle is $4x$ and its length is $6x$. The perimeter of the rectangle is 80. What is the value of x?

    A. 4

    B. 5

    C. 6

    D. 10

30) Jason is 9 miles ahead of Joe running at 5.5 miles per hour and Joe is running at the speed of 7 miles per hour. How long does it take Joe to catch Jason?

    A. 3 hours

    B. 4 hours

    C. 6 hours

    D. 8 hours

31) [6 × (−24) + 8] − (−4) + [4 × 5] ÷ 2 = ?

    A. 148

    B. 132

    C. −144

    D. −122

32) If 40 % of a class are girls, and 25 % of girls play tennis, what percent of the class play tennis?

    A. 10 %

    B. 15%

    C. 20 %

    D. 40 %

33) In a class, there are twice as many girls as boys. If the total number of students in the class is 48, how many girls are in the class?

    A. 16

    B. 24

    C. 32

    D. 36

34) At a Zoo, the ratio of lions to tigers is 5 to 3. Which of the following could NOT be the total number of lions and tigers in the zoo?

    A. 64

    B. 80

    C. 98

    D. 104

35) The price of a sofa is decreased by 25% to $420. What was its original price?

    A. $480

    B. $520

    C. $560

    D. $600

36) A shaft rotates 300 times in 8 seconds. How many times does it rotate in 12 seconds?

    A. 450

    B. 300

    C. 200

    D. 100

37) Solving the equation: $10x - 15.5 = -45.5$?

    A. -3

    B. -2

    C. 2

    D. 3

38) A swimming pool holds 2,000 cubic feet of water. The swimming pool is 25 feet long and 10 feet wide. How deep is the swimming pool?

    A. 2 feet

    B. 4 feet

    C. 6 feet

    D. 8 feet

39) Solve the following equation?

$$6^x = 1,296$$

    A. 3

    B. 4

    C. 5

    D. 6

40) What is the area of the trapezoid?

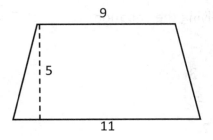

    A. 25

    B. 50

    C. 100

    D. 200

41) $12.124 \div 0.002$?

    A. 6.062

    B. 60.62

    C. 606.2

    D. 6,062

42) What is the value of x in the following equation?

$$10 + 4(x + 5 - 5x) = 30$$

    A. 0

    B. 1

    C. 2

    D. 3

43) Ella bought a pair of gloves for $12.49. She gave the clerk $18.00. How much change should she get back?

    A. $4.51

    B. $5.51

    C. $6.51

    D. $7.51

44) If 60 % of A is 20 % of B, then B is what percent of A?

    A. 3%

    B. 30%

    C. 200%

    D. 300%

45) A card is drawn at random from a standard 52–card deck, what is the probability that the card is of Hearts? (The deck includes 13 of each suit clubs, diamonds, hearts, and spades)

    A. $\frac{1}{3}$

    B. $\frac{1}{4}$

    C. $\frac{1}{6}$

    D. $\frac{1}{52}$

46) $\dfrac{3}{4} + \dfrac{\frac{-2}{5}}{\frac{4}{10}} = ?$

  A. $\dfrac{1}{4}$

  B. $\dfrac{1}{2}$

  C. $-\dfrac{1}{4}$

  D. $-\dfrac{1}{2}$

47) $\dfrac{7 \times 12}{80}$ is closest estimate to?

  A. 1.01

  B. 1.1

  C. 1.3

  D. 1.4

IF YOU FINISH BEFORE TIME IS CALLED, YOU MAY CHECK YOUR WORK ON THIS SECTION.   STOP

ISEE Middle Level

Practice Test 2

Quantitative Reasoning

- o **37 questions**
- o **Total time for this section:** 35 Minutes
- o **Calculators are not allowed at the test.**

1) If the ratio of home fans to visiting fans in a crowd is 3:2 and all 25,000 seats in a stadium are filled, how many visiting fans are in attendance?

    A. 100,000

    B. 10,000

    C. 1,000

    D. 100

2) In following shape y equals to?

    A. 128.5°

    B. 51.5°

    C. 38.5°

    D. 90°

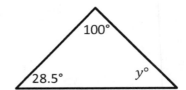

3) Which of the following shows the numbers in increasing order?

    A. $\frac{2}{3}, \frac{8}{11}, \frac{5}{7}, \frac{3}{4}$

    B. $\frac{2}{3}, \frac{5}{7}, \frac{8}{11}, \frac{3}{4}$

    C. $\frac{5}{7}, \frac{3}{4}, \frac{8}{11}, \frac{2}{3}$

    D. $\frac{8}{11}, \frac{3}{4}, \frac{5}{7}, \frac{2}{3}$

4) If an object travels at 0.3 cm per second, how many meters does it travel in 4 hours?

    A. 88.2

    B. 66.4

    C. 50

    D. 43.2

5) What is the value of x in the following equation?

$$\frac{7^x}{49} = 343$$

    A. 5

    B. 4

    C. 3

    D. 2

6) Ava uses a 35% off coupon when buying a sweater that costs $36.12. If she also pays 5% sales tax on the purchase, how much does she pay?

    A. 26.95

    B. 24.65

    C. 21.7

    D. 34.83

7) An item in the store originally priced at $300 was marked down 20%. What is the final sale price of the item?

   A. $240

   B. $204

   C. $195

   D. $200

8) What's the circumference of a circle that has a diameter of 15m?

   A. 47.124 m

   B. 706.9 m

   C. 94.25 m

   D. 36 m

9) What is the area of the shaded region? (one forth of the circle is shaded)

   Diameter = 12

   A. $6\,\pi$

   B. $7\,\pi$

   C. $8\,\pi$

   D. $9\,\pi$

10) If a car has 80-liter petrol and after one hour driving the car use 6-liter petrol, how much petrol remaining after x-hours?

    A. $6x - 80$

    B. $80 + 6x$

    C. $80 - 6x$

    D. $80 - x$

11) A shirt costing $200 is discounted 15%. After a month, the shirt is discounted another 15%. Which of the following expressions can be used to find the selling price of the shirt

    A. $(200)(0.70)$

    B. $(200) - 200(0.30)$

    C. $(200)(0.15) - (200)(0.15)$

    D. $(200)(0.85)(0.85)$

12) Find $\frac{1}{3}$ of $\frac{2}{5}$ of $\frac{3}{4}$ of 290?

    A. 29

    B. 30

    C. 31

    D. 2

13) If $x \le a$ is the solution of $7 + 2x \le 15$, what is the value of a?

  A. $14x$

  B. 4

  C. -4

  D. $15x$

14) The area of the trapezoid below is 132. What is the value of x?

  A. 8

  B. 9

  C. 10

  D. 11

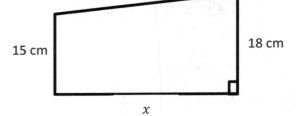

15 cm   18 cm

x

15) Solve for x: $3 + x + 6\left(\frac{x}{2}\right) = 2x + 10$

  A. $\frac{13}{6}$

  B. $\frac{7}{6}$

  C. $\frac{7}{2}$

  D. $\frac{13}{2}$

16) 6 liters of water are poured into an aquarium that's 15cm long, 5cm wide, and 60cm high. How many cm will the water level in the aquarium rise due to this added water? (1 liter of water = 1000 cm³)

    A. 80

    B. 40

    C. 20

    D. 10

17) If $3f + 2g = 3x + y$ and $g = 2y - 3x$, what is f?

    A. $3x + y$

    B. $x + 3y$

    C. $3x - y$

    D. $y - 3x$

18) What is the perimeter of the following parallelogram?

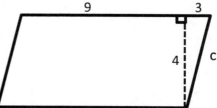

    A. 48

    B. 34

    C. 24

    D. 17

19) In a bundle of 40 fruits, 8 are apples and the rest are bananas. What percent of the bundle is composed of apples?

    A. 47%

    B. 47.75%

    C. 20%

    D. 40.75%

20) What is the value of $\dfrac{-\frac{11}{2}\times\frac{3}{5}}{\frac{11}{30}}$?

    A. -9

    B. 9

    C. $-\frac{1}{9}$

    D. $\frac{1}{9}$

21) The average of 13, 15, 20 and x is 18. What is the value of x?

    A. 9

    B. 15

    C. 18

    D. 24

22) What is the value of mode and median in the following set of numbers?

$$1, 2, 2, 5, 4, 4, 3, 3, 3, 1, 1$$

    A. Mode:1, 2 Median:2

    B. Mode:1, 3 Median:3

    C. Mode:2, 3 Median:2

    D. Mode:1, 3 Median:2.5

23) 5 less than twice a positive integer is 83. What is the integer?

    A. 39

    B. 41

    C. 42

    D. 44

24) If Joe was making $7.50 per hour and got a raise to $7.75 per hour, what percentage increase was the raise?

    A. 2 %

    B. 1.67 %

    C. 3.33 %

    D. 6.66 %

25) Which is the equivalent temperature of 104°F in Celsius? (C = Celsius)

$$C = \frac{5}{9}(F - 32)$$

    A. 32

    B. 38

    C. 40

    D. 44

Quantitative Comparisons

Direction: Questions 26 to 37 are Quantitative Comparisons Questions. Using the information provided in each question, compare the quantity in column A to the quantity in Column B. Choose on your answer sheet grid

A   if the quantity in Column A is greater

B   if the quantity in Column B is greater

C   if the two quantities are equal

D   if the relationship cannot be determined from the information given

26) $2x^5 - 9 = 477$

$\dfrac{1}{3} - \dfrac{y}{5} = -\dfrac{7}{15}$

Quantity A	Quantity B
x	y

27)

Column A	Column B
$4^2 - 2^4$	$2^4 - 4^2$

28) A computer costs $250.

Column A	Column B
A sales tax at 8% of the computer cost	$20

29)

Column A	Column B
$\dfrac{\sqrt{64-48}}{\sqrt{25-9}}$	$\dfrac{(7-4)}{(8-3)}$

30) A

Column A	Column B
The slope of the line $4x + 2y = 7$	The slope of the line that passes through points (2, 5) and (3, 3)

31) The sum of 3 consecutive integers is -45.

Column A	Column B
The largest of these integers	-16

32)

Column A	Column B
$\sqrt{144-81}$	$\sqrt{144} - \sqrt{81}$

33) 6 percent of x is equal to 5 percent of y, where x and y are positive numbers.

Quantity A	Quantity B
x	y

34)

Quantity A	Quantity B
The least prime factor of 55	The least prime factor of 210

35)

Quantity A	Quantity B
$(-5)^4$	5^4

36)

Quantity A	Quantity B
$(1.888)^4(1.888)^8$	$(1.88)^{12}$

37) x is a positive number.

Quantity A	Quantity B
x^{10}	x^{20}

IF YOU FINISH BEFORE TIME IS CALLED, YOU MAY CHECK YOUR WORK ON THIS SECTION ONLY. DO NOT TURN TO ANY OTHER SECTION IN THE TEST.

STOP

www.EffortlessMath.com

229

ISEE Middle Level

Practice Test 2

Mathematics Achievement

- o **47 questions**
- o **Total time for this section:** 40 Minutes
- o **Calculators are not allowed at the test.**

1) If $x =$ lowest common multiple of 30 and 35 then $\frac{x}{2} + 1$ equal to?

    A. 210

    B. 108

    C. 106

    D. 96

2) What is the value of x in the following equation?

$$3^x - 15 = 66$$

    A. 3

    B. 4

    C. 5

    D. 6

3) Which of the following is not synonym for 10^2?

    A. 10 cubed

    B. 10 squared

    C. the square of 10

    D. 10 to the second power

4) If angles A and B are angles of a parallelogram, what is the sum of the measures of the two angles?

    A. 360 degrees
    B. 180 degrees
    C. 90 degrees
    D. Cannot be determined

5) A swing moves from one extreme point (point A) to the opposite extreme point (point B) in 30 seconds. How long does it take that the swing moves 10 times from point A to point B and returns to point A?

    A. 600 seconds
    B. 300 seconds
    C. 200 seconds
    D. 100 seconds

6) There are 2 cars moving in the same direction on a road. A red car is 10 km ahead of a blue car. If the speed of the red car is 50 km per hour and the speed of the blue car is $1\frac{2}{5}$ of the red car, how many minutes will it take the blue car to catch the red car?

    A. 8.5
    B. 15
    C. 30
    D. 50

7) If the area of trapezoid is 100, what is the perimeter of the trapezoid?

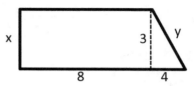

A. 25

B. 35

C. 45

D. 55

8) In two successive years, the population of a town is increased by 15% and 20%. What percent of the population is increased after two years?

A. 32%

B. 35%

C. 38%

D. 42%

9) In 1999, the average worker's income increased $2,000 per year starting from $24,000 annual salary. Which equation represents income greater than average?
(I = income, x = number of years after 1999)

A. $I > 2000x + 24000$

B. $I > -2000x + 24000$

C. $I < -2000x + 24000$

D. $I < 2000x - 24000$

10) Which of the following angles is obtuse?

    A. 20 degrees

    B. 40 degrees

    C. 189 degrees

    D. 110 degrees

11) $5 + 8 \times (-2) - [4 + 22 \times 5] \div 6 = ?$

    A. 120

    B. 88

    C. −30

    D. −20

12) What is ratio of perimeter of figure A to area of figure B?

    A. $\frac{3}{8}$

    B. $\frac{8}{3}$

    C. $\frac{8}{5}$

    D. $\frac{5}{8}$

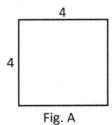

Fig. A

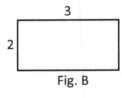

Fig. B

13) Mr. Jones saves $2,500 out of his monthly family income of $55,000. What fractional part of his income does he save?

A. $\frac{1}{22}$

B. $\frac{1}{11}$

C. $\frac{3}{25}$

D. $\frac{2}{15}$

14) Anita's trick–or–treat bag contains 12 pieces of chocolate, 18 suckers, 18 pieces of gum, 24 pieces of licorice. If she randomly pulls a piece of candy from her bag, what is the probability of her pulling out a piece of sucker

A. $\frac{1}{3}$

B. $\frac{1}{4}$

C. $\frac{1}{6}$

D. $\frac{1}{12}$

15) What is the difference in area between a 9 cm by 4 cm rectangle and a circle with diameter of 10 cm? ($\pi = 3$)

A. 49

B. 40

C. 39

D. 6

16) Solve the following equation?

$$(x^2 + 2x + 1) = 100$$

A. $-9, 11$

B. $9, 11$

C. -9

D. 11

17) 120 is equal to?

A. $20 - (4 \times 10) + (6 \times 30)$

B. $\left(\frac{11}{8} \times 72\right) + \left(\frac{125}{5}\right)$

C. $\left(\left(\frac{30}{4} + \frac{13}{2}\right) \times 7\right) - \frac{11}{2} + \frac{110}{4}$

D. $(2 \times 10) + (50 \times 1.5) + 15$

18) $\frac{15 \times 21}{8}$ is closest estimate to?

A. 39.4

B. 39.5

C. 39.6

D. 39.7

19) When a number is multiplied to itself and added by 10, the result is 35. What is the value of the number?

    A. 5 and −5
    B. 6 and −6
    C. 5
    D. 6

20) If you invest $1,000 at an annual rate of 9%, how much interest will you earn after one year?

    A. 9
    B. 9000
    C. 900
    D. 90

21) How many possible outfit combinations come from six shirts, three slacks, and five ties?

    A. 90
    B. 60
    C. 15
    D. 14

22) Solving the equation: $\frac{x}{4} + \frac{5}{4} = \frac{15}{6}$?

    A. 5

    B. 10

    C. 15

    D. 20

23) What is the absolute value of the quantity six minus nine?

    A. -3

    B. 15

    C. -15

    D. 3

24) If $y = 4ab + 3b^3$, what is y when $a = 2$ and $b = 3$?

    A. 24

    B. 105

    C. 51

    D. 36

25) Which of the following angles can represent the three angles of an equilateral triangle?

    A. $10°, 80°, 90°$

    B. $50°, 50°, 80°$

    C. $60°, 60°, 60°$

    D. $45°, 45°, 90°$

26) In the following equation, what is the value of $x - 2y$?

$$x + 3x - 10 = 2\left(\frac{3}{2}x + y\right) - 15$$

    A. 5

    B. -25

    C. -5

    D. 25

27) Two-kilogram apple and three-kilograms orange cost $28 If price of one-kilogram of apple is twice price of one-kilogram of orange. How much does one kilogram apple cost?

    A. $8

    B. $4

    C. $2

    D. $1

28) Which is **not** a prime number?

    A. 181

    B. 151

    C. 131

    D. 121

29) How many tiles of 8 cm² is needed to cover a floor of dimension 6 cm by 24 cm?

    A. 12

    B. 18

    C. 24

    D. 36

30) There are three boxes, a red box, a blue box, and a yellow box. If the weight of the red box is 80 kg and the weight of the red box is 90% of the weight of the blue box, and the weight of the blue box is 120% of the weight of the yellow box, what is the weight of blue and yellow boxes respectively?

    A. 100, 80

    B. 80, 100

    C. 81, 125

    D. 100, 81

31) Each of the x students in a team may invite up to 5 friends to a party. What is the maximum number of students and guests who might attend the party?

    A. $5x + 5$

    B. $5x$

    C. $x + 5$

    D. $6x$

32) Calculate the approximate circumference of the following circle.

    A. 1257

    B. 314

    C. 63

    D. 126

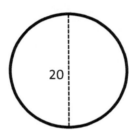

33) 270 minutes=...?

    A. 5 Hours

    B. 4.5 Hours

    C. 3.5 Hours

    D. 0.2 Hours

34) In the figure below, line A is parallel to line B. What is the value of angle x?

A. 35 degree

B. 45 degree

C. 100 degree

D. 145 degree

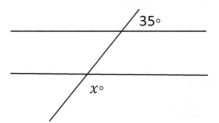

35) $\left(\left((-15) + 40\right) \times \frac{1}{5}\right) + (-10)?$

A. 5

B. 6

C. −5

D. −6

36) You just drove 340 miles and it took you approximately 8 hours. How many miles per hour was your average speed?

A. about 44.5 miles per hour

B. about 42.5 miles per hour

C. about 46.5 miles per hour

D. about 41.5 miles per hour

37) Three people go to a restaurant. Their bill comes to $56.00. They decided to split the cost. One person pays $8.5, the next person pays 2 times that amount. How much will the third person have to pay?

    A. $36.50

    B. $30.50

    C. $41.00

    D. $44.00

38) What is 21,8210 in scientific notation?

    A. 218.21×10^3

    B. 21.821×10^4

    C. 0.21821×10^6

    D. 2.1821×10^5

39) What is the perimeter of the below right triangle?

    A. 60

    B. 30

    C. 20

    D. 10

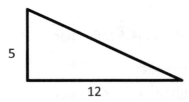

40) If 120 % of a number is equal to 25% of 90, then what is the number?

    A. 19.75

    B. 18.75

    C. 17.75

    D. 16.75

41) What is the value of $(12 - 8)!$?

    A. 20

    B. 24

    C. 4

    D. 28

42) In a department of a company, the ratio of employees with Bachelor's degree to employees with high school Diploma is 2 to 5. If there are 18 employees with Bachelor's degree in this department, how many employees with High School Diploma should be moved to other departments to change the ratio of the number of employees with Bachelor's Degree to the number of employees with High School Diploma to 3 to 4 in this department?

    A. 21

    B. 24

    C. 10

    D. 12

43) The average weight of 18 girls in a class is 60 kg and the average weight of 32 boys in the same class is 62 kg. What is the average weight of all the 50 students in that class?

    A. 61.28

    B. 61.68

    C. 61.90

    D. 62.20

44) What is x in the following right triangle?

    A. $\sqrt{399}$

    B. 20

    C. $\sqrt{401}$

    D. $\sqrt{402}$

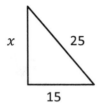

45) John traveled 150 km in 6 hours and Alice traveled 180 km in 4 hours. What is the ratio of the average speed of John to average speed of Alice?

    A. $3 : 2$

    B. $2 : 3$

    C. $5 : 9$

    D. $5 : 6$

46) An angle is equal to one fifth of its supplement. What is the measure of that angle?

A. 20

B. 30

C. 45

D. 60

47) What is the difference of smallest 4–digit number and biggest 4–digit number?

A. 6666

B. 6789

C. 8888

D. 8999

IF YOU FINISH BEFORE TIME IS CALLED, YOU MAY CHECK YOUR WORK ON THIS SECTION.

STOP

ISEE Middle Level Practice Tests
Answers and Explanations

❋ Now, it's time to review your results to see where you went wrong and what areas you need to improve!

ISEE Middle Level Math Practice Test 1 Answer Key								
Quantitative Reasoning					**Mathematics Achievement**			
1	C	17	B	33	D	1	B	17	C	33	A
2	B	18	C	34	A	2	B	18	D	34	C
3	C	19	B	35	C	3	B	19	A	35	C
4	B	20	C	36	C	4	D	20	C	36	A
5	C	21	D	37	A	5	B	21	C	37	A
6	B	22	B			6	B	22	A	38	D
7	B	23	B			7	A	23	D	39	B
8	A	24	A			8	B	24	D	40	B
9	B	25	A			9	C	25	C	41	D
10	D	26	A			10	C	26	D	42	A
11	D	27	B			11	A	27	B	43	B
12	D	28	A			12	C	28	A	44	D
13	C	29	D			13	B	29	A	45	B
14	D	30	A			14	A	30	C	46	C
15	D	31	A			15	D	31	D	47	B
16	D	32	D			16	D	32	A		

ISEE Middle Level Math Practice Test 2 Answer Key

Quantitative Reasoning					Mathematics Achievement				
1	B	17	C	33	B	1	C	17	C	33	B

Quantitative Reasoning:

1	B	17	C	33	B
2	B	18	B	34	A
3	B	19	C	35	C
4	D	20	A	36	C
5	A	21	D	37	D
6	B	22	B		
7	A	23	D		
8	A	24	C		
9	D	25	C		
10	C	26	B		
11	D	27	C		
12	A	28	C		
13	B	29	A		
14	A	30	C		
15	C	31	A		
16	A	32	A		

Mathematics Achievement:

1	C	17	C	33	B
2	B	18	A	34	D
3	A	19	A	35	C
4	D	20	D	36	B
5	A	21	A	37	B
6	C	22	A	38	D
7	B	23	D	39	B
8	C	24	B	40	B
9	A	25	C	41	B
10	D	26	C	42	A
11	C	27	A	43	A
12	B	28	D	44	B
13	A	29	B	45	C
14	B	30	A	46	B
15	C	31	D	47	D
16	B	32	C		

Score Your Test

ISEE scores are broken down by its four sections: Verbal Reasoning, Reading Comprehension, Quantitative Reasoning, and Mathematics Achievement. A sum of the three sections is also reported.

For the Middle Level ISEE, the score range is 760 to 940, the lowest possible score a student can earn is 760 and the highest score is 940 for each section. A student receives 1 point for every correct answer. There is no penalty for wrong or skipped questions.

The total scaled score for a Middle Level ISEE test is the sum of the scores for all sections. A student will also receive a percentile score of between 1-99% that compares that student's test scores with those of other test takers of same grade and gender from the past 3 years.

Use the next table to convert ISEE Middle level raw score to scaled score for application to 7th and 8th grade.

ISEE Middle Level Scaled Scores									
Raw Score	Quantitative Reasoning		Mathematics Achievement		Raw Score	Quantitative Reasoning		Mathematics Achievement	
	7th Grade	8th Grade	7th Grade	8th Grade		7th Grade	8th Grade	7th Grade	8th Grade
0	760	760	760	760	26	900	885	885	865
1	770	765	770	765	27	905	890	885	865
2	780	770	780	770	28	910	895	890	870
3	790	775	790	775	29	910	900	890	870
4	800	780	800	780	30	915	905	895	875
5	810	785	810	785	31	920	910	895	875
6	820	790	820	790	32	925	915	900	880
7	825	795	825	795	33	930	920	900	880
8	830	800	830	800	34	930	925	905	885
9	835	805	835	805	35	935	930	905	885
10	840	810	840	810	36	935	935	910	890
11	845	815	845	815	37	940	940	910	890
12	850	820	850	820	38			915	895
13	855	825	855	825	39			920	900
14	860	830	855	830	40			925	905
15	865	835	860	835	41			925	910
16	870	840	860	840	42			930	915
17	875	845	865	840	43			930	920
18	880	845	865	845	44			935	925
19	880	850	870	845	45			935	930
20	885	855	870	850	46			940	935
21	885	860	875	850	47			940	940
22	890	865	875	855					
23	890	870	875	855					
24	895	875	880	860					
25	895	880	880	860					

ISEE Middle Level Practice Test 1 Answers and Explanations

Quantitative Reasoning

1) **Choice C is correct**

$\frac{1}{6} \cong 0.16$ $\frac{1}{3} \cong 0.33$ $\frac{2}{5} = 0.4$ $\frac{3}{4} = 0.75$

2) **Choice B is correct**

$$452,357,841 \times \frac{1}{10,000} = 45,235.7841$$

3) **Choice C is correct**

Prime factorizing of $36 = 2 \times 2 \times 3 \times 3$

20% off equals $23. Let x be the original price of the table. Then:

$$20\% \ of \ x = 23 \rightarrow 0.2x = 23 \rightarrow x = \frac{23}{0.2} = 115$$

4) **Choice B is correct**

Use the formula for Percent of Change

$$\frac{New \ Value - Old \ Value}{Old \ Value} \times 100 \ \%$$

$\frac{28-40}{40} \times 100 \ \% = -30 \ \%$ (negative sign here means that the new price is less than old price)

5) **Choice C is correct**

$2f = 2 \times (2x - 3y) = 4x - 6y$

$2f + g = 4x - 6y + x + 4y = 5x - 2y$

6) **Choice B is correct**

$$\frac{-48 \times 0.5}{6} = -\frac{48 \times \frac{1}{2}}{6} = -\frac{\frac{48}{2}}{6} = -\frac{48}{12} = -4$$

7) Choice B is correct

Method 1: $8 = 2^3 \rightarrow 8^x = (2^3)^x = 2^{3x}$

$512 = 2^9 \rightarrow 2^{3x} = 2^9 \rightarrow 3x = 9 \rightarrow x = 3$

Method 2: $8^x = 512$

Let's review the choices provided:

    E. 2 $8^x = 512 \rightarrow 8^2 = 64$

    F. 3 $8^x = 512 \rightarrow 8^3 = 512$

    G. 4 $8^x = 512 \rightarrow 8^4 = 4,096$

    H. 5 $8^x = 512 \rightarrow 8^5 = 32,768$

Choice B is correct.

8) Choice A is correct

If the score of Mia was 60, therefore the score of Ava is 30. Since, the score of Emma was half as that of Ava, therefore, the score of Emma is 15.

9) Choice B is correct

Use the formula of areas of circles.

$$Area\ of\ a\ circle = \pi r^2 \Rightarrow 64\,\pi = \pi r^2 \Rightarrow 64 = r^2 \Rightarrow r = 8$$

Radius of the circle is 8. Now, use the circumference formula:

Circumference = 2πr = 2π (8) = 16 π

10) Choice D is correct

$1.12 = \frac{112}{100}$ and $7.2 = \frac{72}{10}$ $\rightarrow 1.12 \times 7.2 = \frac{112}{100} \times \frac{72}{10} = \frac{8064}{1000} = 8.064 \cong 8.1$

11) Choice D is correct

Supplementary angles sum up to 180 degrees. x and 25 degrees are supplementary angles.

Then: $x = 180° - 25° = 155°$

12) Choice D is correct

Let x be the number. Write the equation and solve for x.

$\frac{2}{3} \times 18 = \frac{2}{5} \cdot x \Rightarrow \frac{2 \times 18}{3} = \frac{2x}{5}$, use cross multiplication to solve for x.

$5 \times 36 = 2x \times 3 \Rightarrow 180 = 6x \Rightarrow x = 30$

13) Choice C is correct

Add the first 5 numbers. 40 + 45 + 50 + 35 + 55 = 225

To find the distance traveled in the next 5 hours, multiply the average by number of hours.

Distance = Average × Rate = 50 × 5 = 250

Add both numbers. 250 + 225 = 475

14) Choice D is correct

Mean= $\frac{9+12+29+36+45+63+99+123}{8} = \frac{416}{8} = 52$

15) Choice D is correct

The perimeter of the trapezoid is 54.

Therefore, the missing side (height) is = 54 − 18 − 12 − 14 = 10

 Area of the trapezoid: A = $\frac{1}{2}$ h (b$_1$ + b$_2$) = $\frac{1}{2}$ (10) (12 + 14) = 130

16) Choice D is correct

Let x be the original price. If the price of a laptop is decreased by 10% to $360, then: 90 % of

$x = 360 \Rightarrow 0.90x = 360 \Rightarrow x = 360 \div 0.90 = 400$

17) Choice B is correct

$\frac{2}{5}$ of 120= $\frac{2}{5} \times 120 = 48$

$\frac{1}{4}$ of 48= $\frac{1}{4} \times 48 = 12$

18) Choice C is correct

The ratio of boy to girls is 4:7. Therefore, there are 4 boys out of 11 students. To find the answer, first divide the total number of students by 11, then multiply the result by 4.

$$44 \div 11 = 4 \Rightarrow 4 \times 4 = 16$$

There are 16 boys and 28 (44 – 16) girls. So, 12 more boys should be enrolled to make the ratio 1:1

19) Choice B is correct

Simplify: $4(x + 1) = 6(x - 4) + 20$

$4x + 4 = 6x - 24 + 20,$ $4x + 4 = 6x - 4$

Subtract $4x$ from both sides: $4 = 2x - 4$

Add 4 to both sides: $8 = 2x,\ 4 = x$

20) Choice C is correct

Let x be the sales profit. Then, 2% of sales profit is $0.02x$. Employee's revenue: $0.2x + 7,000$

21) Choice D is correct

Let's review the choices:

A. $\frac{3}{4} > 0.8$ This is not a correct statement. Because $\frac{3}{4} = 0.75$ and it's less than 0.8.

B. $10\% = \frac{2}{5}$ This is not a correct statement. Because 10% = 0.1 and $\frac{2}{5} = 0.4$

C. $3 < \frac{5}{2}$ This is not a correct statement. Because $\frac{5}{2} = 2.5$ and it's less than 3.

D. $\frac{5}{6} > 0.8$ This is a correct statement.

$$\frac{5}{6} = 0.83 \rightarrow 0.8 < \frac{5}{6}$$

22) Choice B is correct

$x = 20 + 125 = 145$

23) Choice B is correct

The diagonal of the square is 8. Let x be the side.

Use Pythagorean Theorem: $a^2 + b^2 = c^2$

$x^2 + x^2 = 8^2 \Rightarrow 2x^2 = 8^2 \Rightarrow 2x^2 = 64 \Rightarrow x^2 = 32 \Rightarrow x = \sqrt{32}$

The area of the square is:

$\sqrt{32} \times \sqrt{32} = 32$

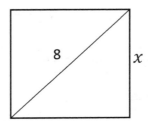

24) Choice A is correct

$-26 - (-65) = -26 + 65 = 65 - 26 = 39$

25) Choice A is correct

Petrol of car A in 250km= $\frac{8 \times 250}{100} = 20$

Petrol of car A in 250km= $\frac{6 \times 250}{100} = 15$, $20 - 15 = 5$

26) Choice A is correct.

Column A: Use order of operation to calculate the result.

$$6 + 4 \times 7 + 8 = 6 + 28 + 8 = 42$$

Column B:

$$4 + 6 \times 7 - 8 \rightarrow 4 + 42 - 8 = 38$$

27) Choice B is correct.

Column A: The value of x when $y = 12$:

$$y = -4x - 8 \rightarrow 12 = -4x - 8 \rightarrow -4x = 20 \rightarrow x = -5$$

Column B: -4

-4 is greater than -5.

28) Choice A is correct.

Column A: Simplify.

$$\sqrt{36} + \sqrt{36} = 6 + 6 = 12$$

12 is greater than $\sqrt{72}$. ($\sqrt{144} = 12$)

29) Choice D is correct.

Column A: Based on information provided, we cannot find the average age of Joe and Michelle or average age of Michelle and Nicole.

30) Choice A is correct.

Column A: Simplify.

$$\sqrt{121 - 64} = \sqrt{57}$$

Column B:

$$\sqrt{121} - \sqrt{64} = 11 - 8 = 3$$

$\sqrt{57}$ is bigger than 3. ($\sqrt{9} = 3$)

31) Choice A is correct

Volume of a right cylinder = $\pi r^2 h \rightarrow 50\pi = \pi r^2 h = \pi(2)^2 h \rightarrow h = 12.5$

The height of the cylinder is 12.5 inches which is bigger than 10 inches.

32) Choice D is correct.

Choose different values for x and find the value of quantity A.

$x = 1$, then:

Quantity A: $\frac{1}{x} + x = \frac{1}{1} + 1 = 2$

Quantity B is greater

$x = 0.1$, then:

Quantity A: $\frac{1}{x} + x = \frac{1}{0.1} + 1 = 10 + 1 = 11$

Quantity A is greater

The relationship cannot be determined from the information given.

33) Choice D is correct.

Simply change the fractions to decimals.

$$\frac{4}{5} = 0.80$$

$$\frac{6}{7} = 0.857 \dots$$

$$\frac{5}{6} = 0.8333 \dots$$

As you can see, x lies between 0.80 and 0.857... and it can be 0.81 or 0.84. The first one is less than 0.833... and the second one is greater than 0.833... .

The relationship cannot be determined from the information given.

34) Choice A is correct.

Simplify quantity B. **Quantity B:** $\left(\frac{x}{6}\right)^6 = \frac{x^6}{6^6}$

Since, the two quantities have the same numerator (x^6) and the denominator in quantity B is bigger $(6^6 > 6)$, then the quantity A is greater.

35) Choice C is correct

Quantity A is: $\frac{3+4+x}{3} = 3 \rightarrow x = 2$

Quantity B is: $\frac{2+(2-6)+(2+4)+(2\times2)}{4} = 2$

36) Choice C is correct

Choose different values for a and b and find the values of quantity A and quantity B.

$a = 2$ and $b = 3$, then:

Quantity A: $|2 - 3| = |-1| = 1$

Quantity B: $|3 - 2| = |1| = 1$

The two quantities are equal.

$a = -3$ and $b = 2$, then:

Quantity A: $|-3 - 2| = |-5| = 5$

Quantity B: $|2 - (-3)| = |2 + 3| = 5$

The two quantities are equal.

Any other values of a and b provide the same answer.

37) Choice A is correct

$$2x^3 + 10 = 64 \rightarrow 2x^3 = 64 - 10 = 54 \rightarrow x^3 = \frac{54}{2} = 27 \rightarrow x = \sqrt[3]{27} = \sqrt[3]{3^3} = 3$$

$$120 - 18y = 84 \rightarrow -18y = 84 - 120 = -36 \rightarrow y = \frac{-36}{-18} = 2$$

ISEE Middle Level Math Practice Test 1 Answers and Explanations

Mathematics Achievement

1) Choice B is correct

15% of 180= $\frac{15}{100} \times 180 = 27$

Let x be the number then, $x = 27 + 12 = 39$

2) Choice B is correct

$\frac{2}{3} \times 90 = 60$

3) Choice B is correct

$$3 \times \left(\frac{1}{3} - \frac{1}{6}\right) + 4 = 3 \times \left(\frac{2-1}{6}\right) + 4 = \frac{3}{6} + 4 = \frac{1}{2} + 4 = \frac{9}{2} = 4.5$$

4) Choice D is correct

Number of pencils are blue= $90 - 43 = 47$

Percent of blue pencils is: $\frac{47}{90} \times 100 = 52.2\% \cong 52\%$

5) Choice B is correct

$(x + 5)^3 = 64 \rightarrow x + 5 = \sqrt[3]{64} = \sqrt[3]{4^3} = 4 \rightarrow x = 4 - 5 = -1$

6) Choice B is correct

50% of 36 is: $\frac{50}{100} \times 36 = \frac{36}{2} = 18$

Let x be the number then: $x = 18 - 5 = 13$

7) Choice A is correct

The perimeter of rectangle is: $2 \times (4 + 7) = 2 \times 11 = 22$

The perimeter of circle is: $2\pi r = 2 \times 3 \times \frac{10}{2} = 30$

Difference in perimeter is: $30 - 22 = 8$

8) Choice B is correct

Use this formula: Percent of Change $= \frac{New\ Value - Old\ Value}{Old\ Value} \times 100\ \%$

$\frac{16000 - 20000}{20000} \times 100\ \% = 20\ \%$ and $\frac{12800 - 16000}{16000} \times 100\ \% = 20\ \%$

9) Choice C is correct

Let x be the number. Write the equation and solve for x.

$(24 - x) \div x = 3$

Multiply both sides by x.

$(24 - x) = 3x$, then add x both sides. $24 = 4x$, now divide both sides by 4. $x = 6$

10) Choice C is correct

$$\left(\left(\frac{3}{2} + 3\right) \times \frac{18}{3}\right) + 63 = \left(\left(\frac{3 + 6}{2}\right) \times \frac{18}{3}\right) + 63 = \left(\frac{9}{2} \times \frac{18}{3}\right) + 63 = 27 + 63 = 90$$

11) Choice A is correct

If $\frac{3x}{2} = 30$, then $3x = 60 \rightarrow x = 20$

$\frac{2x}{5} = \frac{2 \times 20}{5} = \frac{40}{5} = 8$

12) Choice C is correct

$\frac{1}{2} = 0.5 \qquad \frac{7}{9} = 0.77 \qquad 65\% = 0.65$

13) Choice B is correct

Area$= \pi r^2 = \pi \times (\frac{20}{2})^2 = 100\pi = 100 \times 3.14 = 314$

14) Choice A is correct

First, find the number.

Let x be the number. Write the equation and solve for x.

150 % of a number is 75, then: $1.5 \times x = 75 \Rightarrow x = 75 \div 1.5 = 50$

90 % of 50 is: $0.9 \times 50 = 45$

15) Choice D is correct

Find the difference of each pairs of numbers:

2, 3, 5, 8, 12, 17, 23, ___, 38

The difference of 2 and 3 is 1, 3 and 5 is 2, 5 and 8 is 3, 8 and 12 is 4, 12 and 17 is 5, 17 and 23 is 6, 23 and next number should be 7. The number is 23 + 7 = 30

16) Choice D is correct

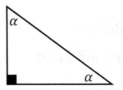

All angles in a triangle sum up to 180 degrees. Then:

$2\alpha + 90° = 180° \rightarrow 2\alpha = 90 \rightarrow \alpha = 45°$

17) Choice C is correct

In rectangle sides that face to face each other is equal.

18) Choice D is correct

$\dfrac{2}{5} \times 25 = \dfrac{50}{5} = 10$

19) Choice A is correct

Let x be the number of shoes the team can purchase. Therefore, the team can purchase 120 x.

The team had $20,000 and spent $14000. Now the team can spend on new shoes $6000 at most. Now, write the inequality: $120x + 14,000 \leq 20,000$

20) Choice C is correct

The capacity of a red box is 20% greater than a blue box. Let x be the capacity of the blue box.

Then: $x + 20\% \ of \ x = 30 \rightarrow 1.2x = 30 \rightarrow x = \dfrac{30}{1.2} = 25$

264

21) Choice C is correct

The question is this: 1.75 is what percent of 1.25?

Use percent formula:

Part $= \frac{\text{percent}}{100} \times$ whole

$1.75 = \frac{\text{percent}}{100} \times 1.25 \Rightarrow 1.75 = \frac{\text{percent} \times 1.25}{100} \Rightarrow 175 = \text{percent} \times 1.25 \Rightarrow \text{percent} = \frac{175}{1.25}$

$= 140$

22) Choice A is correct

60 minutes = 1 Hours $\rightarrow \frac{195}{60} = 3.25$ Hours

23) Choice D is correct

$\frac{2}{5} \times 25 = \frac{50}{5} = 10$

24) Choice D is correct

57 is not prime number, it is divisible by 3.

25) Choice C is correct

The area of the square is 64 inches. Therefore, the side of the square is square root of the area.

$\sqrt{64} = 8$ inches

Four times the side of the square is the perimeter:

4 × 8 = 32 inches

26) Choice D is correct

$12 is what percent of $48?

$12 \div 48 = 0.25 = 25\%$

27) Choice B is correct

Let x be one-kilogram orange cost, then: $3x + (2 \times 4.2) = 26.4 \rightarrow 3x + 8.4 = 26.4 \rightarrow$

$3x = 26.4 - 8.4 \rightarrow 3x = 18 \rightarrow x = \frac{18}{3} = \$6$

28) Choice A is correct

$$\Big(\big((-12) + 20\big) \times 2\Big) + (-15) = \big((8) \times 2\big) - 15 = 16 - 15 = 1$$

29) Choice A is correct

The width of a rectangle is $4x$ and its length is $6x$. Therefore, the perimeter of the rectangle is $20x$. $Perimeter \ of \ a \ rectangle = 2(width + length) = 2(4x + 6x) = 2(10x) = 20x$

The perimeter of the rectangle is 80. Then: $20x = 80 \rightarrow x = 4$

30) Choice C is correct

The distance between Jason and Joe is 9 miles. Jason running at 5.5 miles per hour and Joe is running at the speed of 7 miles per hour. Therefore, every hour the distance is 1.5 miles less.

$9 \div 1.5 = 6$

31) Choice D is correct

Use PEMDAS (order of operation):

$[6 \times (-24) + 8] - (-4) + [4 \times 5] \div 2 = [-144 + 8] - (-4) + [20] \div 2 = [-144 + 8] - (-4) + 10 =$

$[-136] - (-4) + 10 = [-136] + 4 + 10 = -122$

32) Choice A is correct

The percent of girls playing tennis is: $40\ \% \times 25\ \% = 0.40 \times 0.25 = 0.10 = 10\ \%$

33) Choice A is correct

There are twice as many girls as boys. Let x be the number of girls in the class. Then:

$$x + 2x = 48 \rightarrow 3x = 48 \rightarrow x = 16$$

34) Choice C is correct

The ratio of lions to tigers is 5 to 3 at the zoo. Therefore, total number of lions and tigers must be divisible by 8. $5 + 3 = 8$

From the numbers provided, only 98 is not divisible by 8.

35) Choice C is correct

Let x be the original price.

If the price of the sofa is decreased by 25% to $420, then: $75\ \% \ of\ x = 420 \Rightarrow 0.75x = 420 \Rightarrow x = 420 \div 0.75 = 560$

36) Choice A is correct

Number of rotates in 12 second$= \frac{300 \times 12}{8} = 450$

37) Choice A is correct

$$10x = -45.5 + 15.5 = -30 \rightarrow x = \frac{-30}{10} = -3$$

38) Choice D is correct

Use formula of rectangle prism volume.

V = (length) (width) (height) $\Rightarrow$ 2000 = (25) (10) (height) $\Rightarrow$ height = 2000 ÷ 250 = 8

39) Choice B is correct

$1269 = 6^4 \quad \rightarrow 6^x = 6^4 \rightarrow x = 4$

40) Choice B is correct

The area of trapezoid is: $\left(\frac{9+11}{2}\right) \times 5 = 50$

41) Choice D is correct

$$12.124 \div 0.002 = \frac{\frac{12124}{1000}}{\frac{2}{1000}} = \frac{12,124}{2} = 6,062$$

42) Choice A is correct

$$10 + 4(x + 5 - 5x) = 10 + 4(-4x + 5) = 30 \rightarrow 10 - 16x + 20 = 30 \rightarrow -16x + 30 = 30$$
$$\rightarrow -16x = 0 \rightarrow x = 0$$

43) Choice B is correct

$$18 - 12.49 = \$5.51$$

44) Choice D is correct

Write the equation and solve for B:

$0.60A = 0.20B$, divide both sides by 0.20, then you will have $\frac{0.60}{0.20}A = B$, therefore:

$B = 3A$, and B is 3 times of A or it's 300% of A.

45) Choice B is correct

The probability of choosing a Hearts is 13/52 = 1/4

46) Choice C is correct

$$\frac{3}{4} + \frac{\frac{-2}{5}}{\frac{4}{10}} = \frac{3}{4} + \frac{(-2) \times 10}{5 \times 4} = \frac{3}{4} + \frac{-20}{20} = \frac{3}{4} - 1 = \frac{3 - 4}{4} = -\frac{1}{4}$$

47) Choice B is correct

$$\frac{7 \times 12}{80} = \frac{84}{80} = 1.05 \cong 1.1$$

ISEE Middle Level Math Practice Test 2 Answers and Explanations

Quantitative Reasoning

1) Choice B is correct

Number of visiting fans: $\frac{2 \times 25000}{5} = 10,000$

2) Choice B is correct

In triangle sum of all angles equal to 180° then: $y = 180° - (100° + 28.5°) =$

$$100° - 128.5° = 51.5°$$

3) Choice B is correct

$\frac{2}{3} \cong 0.67$ $\frac{5}{7} \cong 0.71$ $\frac{8}{11} \cong 0.73$ $\frac{3}{4} = 0.75$

4) Choice D is correct

One hour equal to 60 minutes then, 4 hours$= 4 \times 60 = 240$ minutes

One minute equal to 60 seconds then, 240 minutes$= 240 \times 60 = 14,400$ seconds

Distance that travel by object is: $0.3 \times 14,400 = 4,320 \, cm = 43.2 \, m$

5) Choice A is correct

$343 = 7^3 \rightarrow \dfrac{7^x}{49} = 7^3 \rightarrow \dfrac{7^x}{7^2} = 7^3 \rightarrow 7^{x-2} = 7^3 \rightarrow x - 2 = 3 \rightarrow x = 5$

6) Choice B is correct

65% of 36.12= $\frac{65}{100} \times 36.12 = \$23.478$

5% of 23.478= $\frac{5}{100} \times 23.478 = \$1.1739$

She pays: $23.478+$1.1739≅$24.65

7) Choice A is correct

20% of 300= $\frac{20}{100} \times 300 = 60$

Final sale price is: $300 - 60 = \$240$

8) Choice A is correct

Circumference of circle= $2\pi r = 2\pi \times \frac{15}{2} = 15\pi \sim 47.124$ m

9) Choice D is correct

Area of circle with diameter 12 is: $\pi r^2 = \pi \left(\frac{12}{2}\right)^2 = 36\pi$

The area of shaded region is: $\frac{36\pi}{4} = 9\pi$

10) Choice C is correct

The amount of petrol consumed after x hours is: $6x$

Petrol remaining: $80 - 6x$

11) Choice D is correct

To find the discount, multiply the number by (100% − rate of discount).

Therefore, for the first discount we get: (200) (100% − 15%) = (200) (0.85) = 170

For the next 15 % discount: (200) (0.85) (0.85)

12) Choice A is correct

$\frac{3}{4}$ of 290= $\frac{3}{4} \times 290 = 217.5$

$\frac{2}{5}$ of 217.5= $\frac{2}{5} \times 217.5 = 87$

$\frac{1}{3}$ of $87=\frac{1}{3} \times 87 = 29$

13) Choice B is correct

$7 + 2x \leq 15 \rightarrow 2x \leq 15 - 7 \rightarrow 2x \leq 8 \rightarrow x \leq \frac{8}{2} \rightarrow x \leq 4$

Then: $a = 4$

14) Choice A is correct

The area of trapezoid is: $\left(\frac{15+18}{2}\right) x = 132 \rightarrow 16.5x = 132 \rightarrow x = 8$

15) Choice C is correct

$$3 + x + 6\left(\frac{x}{2}\right) = 2x + 10 \rightarrow 3 + x + 3x = 2x + 10 \rightarrow 2x = 7 \rightarrow x = 3.5$$

16) Choice A is correct

$One\ liter = 1,000$ cm$^3 \rightarrow 6\ liters = 6,000$ cm^3

$6,000 = 15 \times 5 \times h \rightarrow h = \frac{6000}{75} = 80$ cm

17) Choice C is correct

$$3f + 2g = 3x + y \rightarrow 3f + 2(2y - 3x) = 3x + y \rightarrow 3f + 4y - 6x = 3x + y \rightarrow$$

$$3f = 9x - 3y \rightarrow f = 3x - y$$

18) Choice B is correct

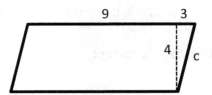

C$= \sqrt{4^2 + 3^2} = \sqrt{25} = 5$

Perimeter of parallelogram$= (9 + 3 + 5) \times 2 = 34$

19) Choice C is correct

$$\frac{8}{40} \times 100 = \frac{8}{4} \times 10 = 20\%$$

20) Choice A is correct

$$\frac{-\frac{11}{2} \times \frac{3}{5}}{\frac{11}{30}} = -\frac{\frac{11 \times 3}{2 \times 5}}{\frac{11}{30}} = -\frac{\frac{33}{10}}{\frac{11}{30}} = -\frac{33 \times 30}{11 \times 10} = -9$$

21) Choice D is correct

$$\text{Average} = \frac{\text{sum of terms}}{\text{number of terms}} \Rightarrow 18 = \frac{13+15+20+x}{4} \Rightarrow 72 = 48 + x \Rightarrow x = 24$$

22) Choice B is correct

We write the numbers in the order: 1, 1, 1, 2, 2, 3, 3, 3, 4, 4, 5

The mode of numbers is: 1 and 3   median is: 3

23) Choice D is correct

Let x be the integer. Then:

$2x - 5 = 83$

Add 5 both sides: $2x = 88$

Divide both sides by 2: $x = 44$

24) Choice C is correct

$$\frac{0.25}{7.5} \times 100 = 3.33$$

25) Choice C is correct

Plug in 104 for F and then solve for C.

$$C = \frac{5}{9}(F - 32) \Rightarrow C = \frac{5}{9}(104 - 32) \Rightarrow C = \frac{5}{9}(72) = 40$$

26) Choice B is correct

$$2x^5 - 9 = 477 \rightarrow 2x^5 = 477 + 9 = 486 \rightarrow x^5 = \frac{486}{2} = 243 \rightarrow x = \sqrt[5]{243} = \sqrt[5]{3^5} = 3$$

$$\frac{1}{3} - \frac{y}{5} = -\frac{7}{15} \rightarrow \frac{y}{5} = \frac{1}{3} + \frac{7}{15} = \frac{5 + 7}{15} = \frac{12}{15} = \frac{4}{5} \rightarrow y = 5 \times \frac{4}{5} = 4$$

27) Choice C is correct.

Column A: $4^2 - 2^4 = 16 - 16 = 0$

Column B: $2^4 - 4^2 = 16 - 16 = 0$

28) Choice C is correct.

Column A: 8% of the computer cost is 20: $8\% \times 250 = 0.08 \times 250 = 20$

Column B: 20

29) Choice A is correct.

Column A: Simplify.

$$\frac{\sqrt{64 - 48}}{\sqrt{25 - 9}} = \frac{\sqrt{16}}{\sqrt{16}} = 1$$

Column B:

$$\frac{(7 - 4)}{(8 - 3)} = \frac{3}{5}$$

30) Choice C is correct.

Column A: The slope of the line $4x + 2y = 7$ is -2.

Write the equation in slope intercept form.

$$4x + 2y = 7 \rightarrow 2y = -4x + 7 \rightarrow y = -2x + \frac{7}{2}$$

Column B: The slope of the line that passes through points (2, 5) and (3, 3):

Use slope formula:

$$slope\ of\ a\ line = \frac{y_2 - y_1}{x_2 - x_1} = \frac{3-5}{3-2} = -2$$

31) Choice A is correct.

Column A: First, find the integers. Let x be the smallest integer. Then the integers are $x, (x + 1)$, and $(x + 2)$. The sum of the integers is -45. Then:

$$x + x + 1 + x + 2 = -45 \rightarrow 3x + 3 = -45 \rightarrow 3x = -48 \rightarrow x = -16$$

Column B: -16

The smallest integer is -16, therefore, the largest integer is bigger than that.

32) Choice A is correct.

Column A: Simplify.

$$\sqrt{144 - 81} = \sqrt{63}$$

Column B:

$$\sqrt{144} - \sqrt{81} = 12 - 9 = 3$$

$\sqrt{63}$ is bigger than 3. ($\sqrt{9} = 3$)

33) Choice B is correct

6% of x = 5% of $y \rightarrow 0.06\ x = 0.05\ y \rightarrow x = \frac{0.05}{0.06}y \rightarrow x = \frac{5}{6}y$, therefore, y is bigger than x.

34) Choice A is correct

prime factoring of 55 is: 5×11

prime factoring of 210 is: $2 \times 3 \times 5 \times 7$

Quantity A = 5 and Quantity B = 2

35) Choice C is correct

Simplify both quantities.

Quantity A: $(-5)^4 = (-5) \times (-5) \times (-5) \times (-5) = 625$

Quantity B: $5 \times 5 \times 5 \times 5 = 625$

The two quantities are equal.

36) Choice C is correct.

Use exponent "product rule": $x^n \times x^m = x^{n+m}$

Quantity A: $(1.888)^4 (1.888)^8 = (1.888)^{4+8} = (1.888)^{12}$

Quantity B: $(1.88)^{12}$

The two quantities are equal.

37) Choice D is correct.

Choose different values for x and find the value of quantity A and quantity B.

$x = 1$, then:

Quantity A: $x^{10} = 1^{10} = 1$

Quantity B: $x^{20} = 1^{20} = 1$

The two quantities are equal.

$x = 2$, then: Quantity A: $x^{10} = 2^{10}$

Quantity B: $x^{20} = 2^{20}$

Quantity B is greater.

Therefore, the relationship cannot be determined from the information given.

ISEE Middle Level Math Practice Test 2 Answers and Explanations

Mathematics Achievement

1) Choice C is correct

Prime factorizing of $30 = 2 \times 3 \times 5$

Prime factorizing of $35 = 5 \times 7$

$x = \text{LCM} = 2 \times 3 \times 5 \times 7 = 210$

$\frac{210}{2} + 1 = 105 + 1 = 106$

2) Choice B is correct

$3^x - 15 = 66 \rightarrow 3^x = 66 + 15 = 81$ and $81 = 3^4$

$3^x = 3^4 \rightarrow x = 4$

3) Choice A is correct

10 cubed is: $10^3 = 1,000$

4) Choice D is correct

All angles in a parallelogram sum up to 360 degrees. Since, we only have 2 angles, therefore the answer cannot be determined.

5) Choice A is correct

Swing moves once from point A to point B and returns to point A is: 30+30=60 seconds

Therefore, for ten times: $10 \times 60 = 600$ seconds

6) Choice C is correct

$$1\frac{2}{5} = \frac{7}{5} = 1.4$$

Speed of the blue car: $\qquad 1.4 \times 50 = 70$

Difference of the cars' speed: $\qquad 70 - 50 = 20$

The red car is 10 km ahead of a blue car. Therefore, it takes 30 minutes to catch the red car. $\frac{10}{20} = 0.5$ Hour= 30 minutes

7) Choice B is correct

The area of trapezoid is: $\left(\frac{8+12}{2}\right) \times x = 100 \rightarrow 10x = 100 \rightarrow x = 10$

Y= $\sqrt{3^2 + 4^2} = 5$

Perimeter is: $12 + 10 + 8 + 5 = 35$

8) Choice C is correct

The population is increased by 15% and 20%. 15% increase changes the population to 115% of original population. For the second increase, multiply the result by 120%.

(1.15) × (1.20) = 1.38 = 138%

38 percent of the population is increased after two years.

9) Choice A is correct

Let x be the number of years. Therefore, $2,000 per year equals $2000x$.

starting from $24,000 annual salary means you should add that amount to $2000x$.

Income more than that is:

I > 2000 x + 24000

10) Choice D is correct

Angle between 90° and 180° is called obtuse angle.

11) Choice C is correct

Use PEMDAS (order of operation):

$5 + 8 \times (-2) - [4 + 22 \times 5] \div 6 = 5 + 8 \times (-2) - [4 + 110] \div 6 = 5 + 8 \times (-2) - [114] \div 6 = 5 + (-16) - 19 = 5 + (-16) - 19 = -11 - 19 = -30$

12) Choice B is correct

Perimeter A $= 4 \times 4 = 16$

Area B $= 2 \times 3 = 6$

$\dfrac{16}{6} = \dfrac{8}{3}$

13) Choice A is correct

2,500 out of 55,000 equals to $\dfrac{2500}{55000} = \dfrac{25}{550} = \dfrac{1}{22}$

14) Choice B is correct

Probability $= \dfrac{number\ of\ desired\ outcomes}{number\ of\ total\ outcomes} = \dfrac{18}{12+18+18+24} = \dfrac{18}{72} = \dfrac{1}{4}$

15) Choice C is correct

The area of rectangle is: $9 \times 4 = 36$ cm^2

The area of circle is: $\pi r^2 = \pi \times (\frac{10}{2})^2 = 3 \times 25 = 75$

Difference in area is: $75 - 36 = 39$

16) Choice B is correct

$x^2 + 2x + 1 = (x + 1)^2 \rightarrow (x + 1)^2 = 100 \rightarrow x + 1 = 10 \rightarrow = 9\ or\ x + 1 = -10 \rightarrow$
$x = -11$

17) Choice C is correct

$$\left(\left(\frac{30}{4}+\frac{13}{2}\right)\times 7\right)-\frac{11}{2}+\frac{110}{4}=\left(\left(\frac{30+26}{4}\right)\times 7\right)-\frac{11}{2}+\frac{55}{2}=\left(\left(\frac{56}{4}\right)\times 7\right)+\frac{55-11}{2}$$

$$=(14\times 7)+\frac{44}{2}=98+22=120$$

18) Choice A is correct

$$\frac{15\times 21}{8}=\frac{315}{8}=39.375\sim 39.4$$

19) Choice A is correct

Let x be the number, then:

$$x^2+10=35\rightarrow x^2=25\rightarrow x^2-25=0\rightarrow (x+5)(x-5)=0\rightarrow x=5\ or\ x=-5$$

20) Choice D is correct

9% of $1000 = $\frac{9}{100}\times 1000=\$90$

21) Choice A is correct

To find the number of possible outfit combinations, multiply number of options for each factor:

6 × 3 × 5 = 90

22) Choice A is correct

$$\frac{x}{4}+\frac{5}{4}=\frac{15}{6}\rightarrow \frac{x}{4}=\frac{15}{6}-\frac{5}{4}=\frac{30-15}{12}=\frac{15}{12}=\frac{5}{4}\rightarrow x=4\times\frac{5}{4}=5$$

23) Choice D is correct

$$|6-9|=|-3|=3$$

24) Choice B is correct

$y = 4ab + 3b^3$

Plug in the values of a and b in the equation: $a = 2$ and $b = 3$

$y = 4 (2) (3) + 3 (3)^3 = 24 + 3(27) = 24 + 81 = 105$

25) Choice C is correct

The angles of an equilateral triangle are 60, 60, 60 degrees.

26) Choice C is correct

$$x + 3x - 10 = \left(2 \times \left(\frac{3}{2} + y\right)\right) - 15 \rightarrow 4x - 10 = \left(2 \times \frac{3}{2}x\right) + 2y - 15 \rightarrow 4x - 10$$
$$= 3x + 2y - 15 \rightarrow 4x - 3x - 2y = 10 - 15 \rightarrow x - 2y = -5$$

27) Choice A is correct

Let x be price of one-kilogram of apple and y be price of one-kilogram of orange, then: $x = 2y$

$$2x + 3y = 28 \rightarrow 2(2y) + 3y = 28 \rightarrow 7y = 28 \rightarrow y = \frac{28}{7} = 4 \rightarrow x = 2 \times 4 = 8$$

28) Choice D is correct

121 is not prime number, it is divided by 11

29) Choice B is correct

The area of the floor is: 6 cm × 24 cm = 144 cm

The number is tiles needed = 144 ÷ 8 = 18

30) Choice A is correct

Weight of blue box= $\frac{10 \times 90}{9} = 100$

Weight of yellow box= $\frac{100 \times 100}{25} = 80$

31) Choice D is correct

Since, each of the x students in a team may invite up to 5 friends, the maximum number of people in the party is 6 times x or $6x$. (one student + 5 friends = 6 people)

32) Choice C is correct

Perimeter= $2\pi r = 2 \times \pi \times \frac{20}{2} = 20 \cong 62.8 \cong 63$

33) Choice B is correct

60 minutes = 1Hours→ $\frac{270}{60} = 4.5$ Hours

34) Choice D is correct

$180° - 35° = 145°$

35) Choice C is correct

$$\left(\left((-15) + 40\right) \times \frac{1}{5}\right) + (-10) = \left((25) \times \frac{1}{5}\right) - 10 = 5 - 10 = -5$$

36) Choice B is correct

Average speed: $\frac{340}{8} = 42.5$ miles per hour

37) Choice B is correct

Let x be the price that third person has to pay then; $56 = 8.5 + (2 \times 8.5) + x \rightarrow$

$$x = 56 - 25.5 = 30.5$$

38) Choice D is correct

$218,210 = 2.1821 \times 10^5$

39) Choice B is correct

c= $\sqrt{5^2 + 12^2} = \sqrt{169} = 13$

Perimeter is: $5 + 12 + 13 = 30$

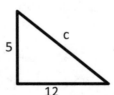

40) Choice B is correct

Let x be the number then, $120\,\%$ of $x = 1.2x$

$$1.2x = 0.25 \times 90 = 22.5 \rightarrow x = \frac{22.5}{1.2} = 18.75$$

41) Choice B is correct

$(12 - 8)! = 4! = 4 \times 3 \times 2 \times 1 = 24$

42) Choice A is correct

Number of employees if ratio is 2 to 5:　$\frac{7 \times 18}{2} = 63$

Number of employees with high school Diploma if ratio is 2 to 5$= 63 - 18 = 45$

Number of employees if ratio is 3 to 4:　$\frac{7 \times 18}{3} = 42$

Number of employees with high school Diploma if ratio is 3 to 4$= 42 - 18 = 24$

Number of employees with High School Diploma should be moved to other departments:

$$45 - 24 = 21$$

43) Choice A is correct

average $= \frac{\text{sum of terms}}{\text{number of terms}}$

The sum of the weight of all girls is: 18 × 60 = 1080 kg

The sum of the weight of all boys is: 32 × 62 = 1984 kg

The sum of the weight of all students is: 1080 + 1984 = 3064 kg

Average $= \frac{3064}{50} = 61.28$

44) Choice B is correct

$x^2 + 15^2 = 25^2 \rightarrow {}^2 = 25^2 - 15^2 \rightarrow x = \sqrt{25^2 - 15^2} \rightarrow x = \sqrt{625 - 225} = \sqrt{400} = 20$

45) Choice C is correct

The average speed of john is: $150 \div 6 = 25$

The average speed of Alice is: $180 \div 4 = 45$

Write the ratio and simplify.

$25 : 45 \Rightarrow 5 : 9$

46) Choice B is correct

The sum of supplement angles is 180. Let x be that angle. Therefore, $x + 5x = 180$

$6x = 180$, divide both sides by 6: $x = 30$

47) Choice D is correct

Smallest 4–digit number is 1000, and biggest 4–digit number is 9999. The difference is: 8999

"Effortless Math" Publications

Effortless Math authors' team strives to prepare and publish the best quality Mathematics learning resources to make learning Math easier for all. We hope that our publications help you or your student Math in an effective way.

We all in Effortless Math wish you good luck and successful studies!

Effortless Math Authors

www.EffortlessMath.com

... So Much More Online!

✓ FREE Math lessons

✓ More Math learning books!

✓ Mathematics Worksheets

✓ Online Math Tutors

Need a PDF version of this book?

Please visit www.EffortlessMath.com